Modeling the Wireless Propagation Channel

Wiley Series on Wireless Communications and Mobile Computing

Series Editors: Dr Xuemin (Sherman) Shen, *University of Waterloo, Canada*
Dr Yi Pan, *Georgia State University, USA*

The "Wiley Series on Wireless Communications and Mobile Computing" is a series of comprehensive, practical and timely books on wireless communication and network systems. The series focuses on topics ranging from wireless communication and coding theory to wireless applications and pervasive computing. The books offer engineers and other technical professionals, researchers, educators, and advanced students in these fields with invaluable insight into the latest developments and cutting-edge research.

Other titles in the series:

Mišić and Mišić: *Wireless Personal Area Networks: Performance, Interconnections and Security with IEEE 802.15.4*, January 2007, 978-0-470-51847-2

Takagi and Walke: *Spectrum Requirement Planning in Wireless Communications: Model and Methodology for IMT-Advanced,* April 2008, 978-0-470-98647-9

Ippolito: *Satellite Communications Systems Engineering: Atmospheric Effects, Satellite Link Design and System Performance*, September 2008, 978-0-470-72527-6

Lin and Sou: *Charging for All-IP Telecommunications*, September 2008, 978-0-470-77565-3

Myung: *Single Carrier FDMA: A New Air Interface for 3GPP Long Term Evolution*, November 2008, 978-0-470-72449-1

Hart, Tao and Zhou: *IEEE 802.16j Mobile Multihop Relay*, March 2009, 978-0-470-99399-6

Qian, Muller and Chen: *Security in Wireless Networks and Systems*, May 2009, 978-0-470-51212-8

Wang, Kondi, Luthra and Ci: *4G Wireless Video Communications*, May 2009, 978-0-470-77307-9

Cai, Shen and Mark: *Multimedia for Wireless Internet–Modeling and Analysis,* May 2009, 978-0-470-77065-8

Stojmenovic: *Wireless Sensor and Actuator Networks: Algorithms and Protocols for Scalable Coordination and Data Communication*, August 2009, 978-0-470-17082-3

Modeling the Wireless Propagation Channel

A Simulation Approach with MATLAB®

F. Pérez Fontán and P. Mariño Espiñeira

University of Vigo, Spain

WILEY

A John Wiley and Sons, Ltd, Publication

This edition first published 2008
© 2008 John Wiley & Sons Ltd

Registered office

John Wiley & Sons Ltd, The Atrium, Southern Gate, Chichester, West Sussex, PO19 8SQ, United Kingdom

For details of our global editorial offices, for customer services and for information about how to apply for permission to reuse the copyright material in this book please see our website at www.wiley.com.

Library of Congress Cataloging-in-Publication Data

Pérez Fontán, F.
    Modeling the wireless propagation channel : a simulation approach with MATLAB® / F. Pérez Fontán and
P. Mariño Espiñeira.
        p. cm.
    Includes bibliographical references and index.
    ISBN 978-0-470-72785-0 (cloth)
  1. Mobile communication systems – Computer simulation. 2. Radio wave propagation – Computer simulation.
3. Radio wave propagation – Mathematical models. 4. Antennas (Electronics) – Computer simulation.
5. MATLAB®. I. Mariño Espiñeira, P. II. Title.
    TK6570.M6P45 2008
    621.384' 11–dc22   2008014495

A catalogue record for this book is available from the British Library.

ISBN 978-0-470-72785-0 (HB)

Typeset in 10/12pt Times by Thomson Digital Noida, India.

For Tere, Nando and, especially, Celia
For Marysol, Roy and Iria

Contents

About the Series Editors

Xuemin (Sherman) Shen (M'97-SM'02) received the B.Sc. degree in electrical engineering from Dalian Maritime University, China in 1982, and the M.Sc. and Ph.D. degrees (both in electrical engineering) from Rutgers University, New Jersey, USA, in 1987 and 1990, respectively. He is a Professor and University Research Chair, and the Associate Chair for Graduate Studies, Department of Electrical and Computer Engineering, University of Waterloo, Canada. His research focuses on mobility and resource management in interconnected wireless/wired networks, UWB wireless communications systems, wireless security, and ad hoc and sensor networks. He is a co-author of three books, and has published more than 300 papers and book chapters in wireless communications and networks, control and filtering. Dr. Shen serves as a Founding Area Editor for *IEEE Transactions on Wireless Communications*; Editor-in-Chief for *Peer-to-Peer Networking and Application*; Associate Editor for *IEEE Transactions on Vehicular Technology*; *KICS/IEEE Journal of Communications and Networks*, *Computer Networks*; *ACM/Wireless Networks*; and *Wireless Communications and Mobile Computing* (Wiley), etc. He has also served as Guest Editor for *IEEE JSAC, IEEE Wireless Communications*, and *IEEE Communications Magazine*. Dr. Shen received the Excellent Graduate Supervision Award in 2006, and the Outstanding Performance Award in 2004 from the University of Waterloo, the Premier's Research Excellence Award (PREA) in 2003 from the Province of Ontario, Canada, and the Distinguished Performance Award in 2002 from the Faculty of Engineering, University of Waterloo. Dr. Shen is a registered Professional Engineer of Ontario, Canada.

Dr. Yi Pan is the Chair and a Professor in the Department of Computer Science at Georgia State University, USA. Dr. Pan received his B.Eng. and M.Eng. degrees in computer engineering from Tsinghua University, China, in 1982 and 1984, respectively, and his Ph.D. degree in computer science from the University of Pittsburgh, USA, in 1991. Dr. Pan's research interests include parallel and distributed computing, optical networks, wireless networks, and bioinformatics. Dr. Pan has published more than 100 journal papers with over 30 papers published in various IEEE journals. In addition, he has published over 130 papers in refereed conferences (including IPDPS, ICPP, ICDCS, INFOCOM, and GLOBECOM). He has also co-edited over 30 books. Dr. Pan has served as an editor-in-chief or an editorial board member for 15 journals including five *IEEE Transactions* and has organized many international conferences and workshops. Dr. Pan has delivered over 10 keynote speeches at many international conferences. Dr. Pan is an IEEE Distinguished Speaker (2000–2002), a Yamacraw Distinguished Speaker (2002), and a Shell Oil Colloquium Speaker (2002). He is listed in *Men of Achievement, Who's Who in America, Who's Who in American Education, Who's Who in Computational Science and Engineering,* and *Who's Who of Asian Americans.*

Preface

This book deals with several issues related to the wireless propagation channel using a simulation approach. This means that we will be generating synthetic, but realistic, series of relevant propagation parameters as a function of time or traversed distance. It is hoped that this will allow the reader to become acquainted with this topic in a simpler, more intuitive way than by using cumbersome mathematical formulations. This will be done in a step-by-step fashion whereby new concepts will be presented as we progress in our familiarization process with the wireless propagation channel.

This book will reproduce the characteristics of the wanted and interfering signals and especially their dynamics. Thus, time series will be generated in most of the projects proposed. Sometimes, the term 'time series' will be used indistinctly for series with the abscissas in time units and in length units (traveled distance), thus explicitly showing the time and/or location variability in the received signal, or in the propagation channel parameters of interest: power, voltage, phase, attenuation, frequency response, impulse response, etc. The procedure followed in all chapters will be to first define a propagation scenario, either statistically or geometrically, then generate one or several time series and, finally, perform a statistical analysis of the produced series, in order to extract relevant parameters from which conclusions may be drawn. Only in Chapter 1 we will be using series already produced using simulators from other chapters.

Part of the proposed material currently complements a lecture on Mobile Communications Systems corresponding to the fifth year of a Telecommunications Engineering course (Masters degree) at the University of Vigo, Spain. This lecture is complemented by simulation work carried out during one semester in sessions of two hours per week, where several issues related to the mobile radio channel and, in general, to any radio channel are studied by means of simulations carried out in MATLAB® (MATLAB® is a registered trademark of The MathWorks, Inc.).

This book covers a number of issues related to the mobile or wireless propagation channel on a step-by-step basis, where we briefly introduce the needed theoretical background relevant to the topic at the beginning of each chapter. This book is not intended as a theoretical book with a full coverage of the various fundamental concepts, and existing models. It is rather a complement to such books. A thorough coverage of the various topics discussed in this book can found in some of the books mentioned in the References of the various chapters.

This book is structured as follows: first a brief theoretical background is provided, then a number of simulation projects are proposed. In some cases, further theoretical background relative to each specific project is provided. Then, we go on to describe the simulation approach followed. The basic result in each simulation is a synthetic series. This is further processed to obtain a number of statistical parameters, e.g., a cumulative distribution. The obtained results are commented upon to point out the underlying concepts, and the influence of the inputs employed. The codes for all projects will be provided by the authors on request, see email addresses below. The reader is also prompted to change the inputs to observe the influence of such changes in the obtained results. Sometimes, the reader is also encouraged to modify the code provided to implement other algorithms not included in the text. Each chapter ends with a list of references, and the list of MATLAB® scripts and functions used.

The MATLAB® software developed is not totally optimized, since the main objective is that the algorithms implemented can be better understood by the reader. Such programs can also be modified by the reader and used in the development of more complex simulators.

If the reader identifies problems in the software, he or she is encouraged to report them to the authors (fpfontan@tsc.uvigo.es or pmarino@uvigo.es) so that we can make the required updates or provide solutions. Readers can also contact the above email addresses in case the web page address (www.dte.uvigo.es/modeling_wiley) for downloading the software mentioned in the book is changed. We have used *version 7 of MATLAB®*, and the programs may not work on earlier versions. MATLAB®'s *signal processing toolbox* should also be available for a proper operation of the programs supplied.

We recommend this book for different types of readers going through different stages in their process to approaching the wireless propagation channel. Possible readers include senior undergraduates, and first-year graduate instructors and students for complementing their lecture materials with a practical, simulation-based book. The MATLAB® functions and scripts provided should allow them to quickly step through the first learning phases. The programs provided could help these students in developing their own programs, for example, for their project and thesis work. The material supplied could also be of interest to first-year Ph.D. students.

Other potential readers could be the self-learning professionals who need to get acquainted with this topic by first gaining a good intuitive insight into the wireless channel. In a second step, these readers could go on and approach other texts where more thorough mathematical and statistical developments of the various phenomena are available.

Other possible readers are practicing engineers who are involved in projects where the dynamics of the channel need to be included in their studies. System engineers who work on the definition of the transmission scheme needed for a new system under development, where a good understanding of the basic propagation phenomena is needed, can find the help they may require on the channel dynamics in this book

This book is structured into the following chapters. In Chapter 1, *Introduction to Wireless Propagation*, we present a brief introduction to the basic concepts and mechanisms driving the wireless propagation channel. Then, we present very simple time-series analysis techniques which cover the basics previously introduced. These include the fast variations due to multipath and the combined effects of shadowing and multipath. Finally, we have a look into the signal's complex envelope, and we plot its magnitude and phase. We also present the Rayleigh and Rice cases which correspond to harsh and benign propagation conditions.

In Chapter 2, *Shadowing Effects*, we present several techniques for quantifying the shadowing effects caused by totally absorbing screens. Even though the formulations presented are somehow more involved than those in other chapters, we have thought it helpful to provide the reader with tools to reproduce the effect of the presence of buildings and other large obstacles. In this way, the reader will be able to replicate the effects of the terrain or the signal variations while driving along a street. We also wanted to point out that, through fairly simple physical models, it is possible to reproduce the inherent cross-correlation properties observed between wanted or interfering signals converging on a given mobile terminal.

Chapter 3, *Coverage and Interference*, looks into the shadowing phenomenon from a statistical point of view. In Chapter 2 we looked at it from a deterministic point of view. We provide a number of simulation cases where the various aspects of shadowing are presented, thus, we first look at the normal distribution describing the variability of shadowing effects when expressed in dB, and we link the mean of the distribution to the path loss. We define two additional concepts, the location variability and the correlation length. We present alternative ways of reproducing such variations. We then present the effect of cross-correlation, and how to introduce it in simulated series using a statistical approach. Finally, we introduce the multiple interference case. We also use the generated series in hard- and soft-handover examples.

In Chapter 4, *Introduction to Multipath*, after studying the shadowing effects, we start the analysis of the multipath phenomenon. This is introduced by presenting first, very simple geometries, and continuing with more intricate ones. We point out how multipath is a *time-selective* phenomenon, i.e., it gives rise to fades. In addition, the movement of at least one of the terminals causes Doppler effects. We see how the *Doppler shift* and the angle of arrival of a given echo are interrelated, and that there is a limit to the maximum rate of change possible, if it is only the terminal that is moving. Throughout, we assume that multipath gives rise to a spatial standing wave, sensed by the terminal antenna as it moves. We also show a possible way of generating time variations when the mobile terminal is stationary. Finally, we briefly introduce the case where both terminals, and even the scatterers, are moving. In this chapter, and throughout this book, we use a very simple approach for the multipath channel based on the so-called *multiple point-scatterer model*.

In Chapter 5, *Multipath: Narrowband Channel*, we continue our discussion on the narrowband multipath channel. In this case, we go one step further and use normalized levels so that we are able to introduce our synthetic time series in actual link-level simulators, where the working point (the average signal-to-noise ratio) is extremely important. We continue our discussion using the multiple point-scatterer model to simulate second-order statistics, and other parameters. We then introduce in the model a direct ray, thus generating a Rice series. Afterward, we present alternative ways for generating Rayleigh and Rice series, one consisting in an array of low frequency sinusoidal generators, and the other consisting on the combination in quadrature of two random Gaussian, noise-like signals, which are also filtered to force the wanted Doppler characteristics. Finally, we look into the issue of space diversity, both at the mobile and the base station sides. The concept of diversity gain/improvement is introduced.

In Chapter 6, *Shadowing and Multipath*, we conclude our analysis of the narrowband channel by modeling together the slow and fast variations due to shadowing and multipath, respectively. We present two mixed distributions: Suzuki and Loo. We also simulate a very simple power control technique.

Chapter 7, *Multipath: Wideband Channel*, completes our picture of the channel. When the time spreading is significant with respect to the duration of the transmitted symbols, selective fading starts to be important. We introduce techniques for characterizing wideband channels, starting with deterministic time-varying channels, and then going on to introduce the stochastic characterization. The main functions and parameters necessary are discussed. We put the wideband channel in the frame of the multiple point-scatterer model, which has been used for presenting illustrative channel simulations. Finally, a statistical model, that of COST 207, is implemented.

In Chapter 8, *Propagation in Microcells and Picocells*, we review several issues of relevance for the microcell and picocell scenarios. We also consider the outdoor-to-indoor propagation case. In addition to reviewing some basic theoretical and empirical techniques, we propose and implement simulations which deal with the modeling of these scenarios using simple, image-theory ray-tracing techniques, which reproduce what empirical models forecast. Moreover, we introduce the consideration of diffraction effects in these scenarios. Finally, we also present a widespread statistical model due to Saleh and Valenzuela for describing the wideband indoor channel, where we present the concept of ray clustering.

In Chapter 9, *The Land Mobile Satellite Channel*, we become acquainted with some of the issues specific to this channel, and its more common modeling techniques. We present alternative ways for generating time series, i.e., using a fully statistical approach based on Markov chains, and a mixed statistical-deterministic approach, called here the *virtual city* approach. Additionally, we become familiar with simple techniques for assessing multiple satellite availability by making use of *constellation simulator* data, together with *urban masks*. Finally, we quantify the Doppler shift caused by non-GEO satellites.

Finally, Chapter 10, *The Directional Wireless Channel*, concludes our study on the wireless channel. Here we study the spatial properties of the multipath channel. We first learn how scatterer contributions tend to be clustered in terms of excess delays, which indicate that they can belong to the same obstacle. We also see that these clustered contributions are spread out in angle of arrival and departure. We also simulate the multiple antenna channel (MIMO, *multiple-input multiple-output*) using our point-scatterer approach and we show how the capacity can be increased substantially. Finally, we present another approach, a statistical one, for simulating the MIMO channel.

There are many other simulations that could have been proposed to the reader. It is hoped, however, that the available ones can be used as a starting point for becoming familiar with most of the channel features discussed.

www.dte.uvigo.es/modeling_wiley

Acknowledgments

We would like to specially thank Professor José María Hernando for the opportunities provided through the years, especially working with and learning from him. To Bertram Arbesser-Rastburg and Pedro Baptista, the use of the multiple point-scatterer model comes from that time. Also thanks to two great colleagues and friends, Erwin Kubista and Uwe Fiebig. Finally, thanks to our colleagues here at the University of Vigo, especially Iria Sanchez and Belén Sanmartín who did a lot of the programming and reviewing, but also to our other colleagues Adolfo (Fito) Nuñez, Pavel Valtr, Ana Castro, Fernando Machado and Vicente Pastoriza.

1

Introduction to Wireless Propagation

1.1 Introduction

This book deals with several topics related to mobile and, in general, wireless propagation channels using a simulation approach. This means that we will be generating synthetic, but realistic, series of relevant propagation parameters as a function of time or traversed distance. It is hoped that this will allow the reader to become acquainted with this topic in a simpler, more intuitive way than by using cumbersome mathematical formulations.

Typically, frequencies in the VHF/UHF bands, and slightly above, are used for wireless applications. A number of propagation mechanisms such as reflections, diffractions, transmissions, etc. dominate. These effects are normally caused by environmental features close to the user terminal or mobile station, MS. In some cases, also the other end of the link, the base station, BS, or network access point will be surrounded by local features affecting the propagation characteristics. Furthermore, in some cases, far-off, large environment elements such as mountains or buildings may also intervene in the link characteristics, causing significant time spreading.

The frequency bands mentioned above are well suited for area coverage, including outdoor-to-indoor and indoor-to-indoor links. Similar frequencies may be used in fixed local access systems (point-to-point and point-to-multipoint) where identical effects can be expected, the main difference being that the channel variability and time spreading will be much smaller.

The wireless channel, in the same way as the various wired channels (optical fiber, coaxial, waveguides, twisted pair, power line, etc.), should provide a distortion-free link between the transmitter and the receiver. This is achieved if the magnitude of its frequency response is flat and the phase is linear, i.e.,

$$|H(f)| = \text{constant and } \arg(H(f))\alpha f \tag{1.1}$$

Modeling the Wireless Propagation Channel   F. Pérez Fontán and P. Mariño Espiñeira
© 2008 John Wiley & Sons, Ltd

where α means proportional to. Equivalently, in the time domain, the associated impulse response should be of the form

$$h(t) = a \exp(\mathrm{j}\xi)\delta(t - \tau), \tag{1.2}$$

where ξ is the phase. The following Fourier transform pair holds,

$$h(t) = a \exp(\mathrm{j}\xi)\delta(t - \tau) \overset{F}{\longleftrightarrow} H(f) = a \exp(\mathrm{j}\xi) \exp(-\mathrm{j}\omega\tau) \tag{1.3}$$

where the specifications for a distortion-free channel are fulfilled.

Signals transmitted through the radio channel use a limited portion of the spectrum, small in comparison with the central frequency or the carrier, f_c. Such signals are called pass-band signals and follow the expression

$$y(t) = a(t) \cos[2\pi f_c t + \theta(t)] \tag{1.4}$$

where $a(t)$ is the envelope of $y(t)$ and $\theta(t)$ is the phase. As all the transmitted information (modulation) is contained in the envelope and the phase, the signal can be analyzed by just using the so-called complex envelope or low-pass equivalent of $y(t)$, given by

$$r(t) = a(t) \exp[\mathrm{j}\theta(t)] \tag{1.5}$$

It will always be possible to recover $y(t)$ from $r(t)$ by simple multiplication by the carrier phasor and taking the real part, i.e.,

$$y(t) = \mathrm{Re}[r(t) \exp(\mathrm{j}2\pi f_c t)] \tag{1.6}$$

Figure 1.1 shows how the complex envelope can be regarded as a version of the pass-band (RF) signal shifted to a central frequency equal to 0 Hz. The spectrum of the pass-band signal, $Y(f)$, can be put in terms of the complex envelope spectra, $R(f)$, using the expression [1]

$$Y(f) = \frac{1}{2}[R(f - f_c) + R^*(-f - f_c)] \tag{1.7}$$

where $R(f)$ is the Fourier transform of $r(t)$, and * indicates complex conjugate.

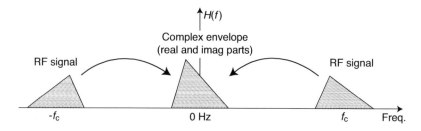

Figure 1.1   Signal spectrum in the band-pass and low-pass domains

The complex envelope, $r(t)$, instead of being represented in terms of its magnitude and phase, can also be put in terms of its in-phase and quadrature components, i.e.,

$$r(t) = a(t)\exp[j\theta(t)] = I(t) + jQ(t) \tag{1.8}$$

In our simulations we will always work in the low-pass equivalent domain.

The complex envelope is in voltage units. To compute the instantaneous power, its magnitude has to be squared, i.e.,

$$p(t) = \frac{1}{2}\frac{|r(t)|^2}{R} \tag{1.9}$$

where R is the resistance of the load.

We can try and find the average received power when an unmodulated, continuous wave (CW) is transmitted. The squared rms value of the RF signal voltage can be linked to the rms squared value of the magnitude of complex envelope, thus [1]

$$y_{rms}^2 = \overline{y^2(t)} = \frac{1}{2}\overline{|r(t)|^2} = \frac{1}{2}r_{rms}^2 \tag{1.10}$$

where $\bar{x}$ (overbar) means time average of x. The average power is given by

$$\bar{p} = \frac{y_{rms}^2}{R} = \frac{\overline{y^2(t)}}{R} = \frac{\overline{|r(t)|^2}}{2R} = \frac{r_{rms}^2}{2R} \tag{1.11}$$

where $\bar{p}$ is the average power. We will be making reference later in this chapter to the rms value of the complex envelope for the Rayleigh case (Project 1.1).

Unfortunately, the channel response does not remain constant in time as the terminal moves. Even when the terminal is stationary, time variations may arise. The channel can also introduce distortion, frequency shifts and other effects as we will be discussing in this and other chapters.

Furthermore, the frequency spectrum is not used exclusively by the wanted link: adjacent frequencies, or even the same frequency, are used in an ever increasingly crowded spectrum. Thus, it is important to perform a joint analysis of the wanted link and interfering links. This point will be discussed in some detail in Chapter 3.

This book will be reproducing the characteristics of the wanted and interfering signals, especially their dynamics. Thus, time series will be generated in most of the projects proposed throughout. Sometimes, the term 'time series' will be used indistinctly for 'series' with the abscissas in time units or in length units (traveled distance), thus explicitly showing the time and/or location variability in the received signal or in the propagation channel parameters of interest: power, voltage, phase, attenuation, frequency response, etc.

The procedure to be followed in all chapters will be to first define a propagation scenario, either statistically or geometrically, then generate one or several time series and, finally, perform a statistical analysis of the produced time series for extracting relevant parameters from which conclusions may be drawn. The only exception is in this chapter where we will be using series already produced using simulators from other chapters. We will be analyzing these series in order to get acquainted with some of the basic techniques that will be used throughout the book.

1.2 Wireless Propagation Basics

Now, we would like to briefly discuss some of the characteristics of the mobile/wireless propagation channel [2]. The modeling techniques involved in land mobile systems have also many similarities with those used in other *area coverage systems* such as sound and TV broadcasting. The convergence of two-way and broadcast, as well as of mobile and fixed systems is making such systems almost indistinguishable, as they try to aim at the same users with offerings of similar services. The similarities between fixed and mobile wireless channels over the frequency bands of interest not only include the mechanisms giving rise to path loss, but also they are subjected to shadowing and multipath effects even though these are normally much milder.

Depending on the location and the BS or access point height, cells of larger or smaller size can be created. The classical cellular environment of tall masts above rooftops gives rise to so-called *macrocells*. Propagation in these conditions will take up most of the discussion in this chapter and most of this book. As the BS antenna height becomes smaller and is below the surrounding rooftops, so-called *microcells* are generated. BSs within buildings give rise to *picocells* (Chapter 8). However, when satellites are used, which means much higher 'BS antenna heights', *megacells* are originated (Chapter 9).

Man-made structures [3] such as buildings or small houses in suburban areas, with sizes ranging from a few meters to tens of meters, dramatically influence the wireless propagation channel. In urban areas, the size of structures can be even larger. Likewise, in rural and suburban environments, features such as isolated trees or groups of trees, etc. may reach similar dimensions. These features are similar or greater in size than the transmitted wavelength (*metric, decimetric, centimetric waves*) and may both block and scatter the radio signal causing specular and/or diffuse reflections. These contributions may reach MS by way of multiple paths, in addition to that of the direct signal. In many cases, these echoes make it possible that a sufficient amount of energy reaches the receiver, so that the communication link is feasible. This is especially so when the direct signal is blocked. Hence, in addition to the expected distance power decay, two main effects are characteristic in mobile propagation: *shadowing* and *multipath*.

We can identify three levels in the rate of change of the received signal as a function of the distance between BS and MS, namely, *very slow variations* due to range, *slow* or *long-term variations* due to shadowing and *fast* or *short-term variations* due to multipath.

While in conventional macrocells, BS heights are in the order of 30 m and are normally set on elevated sites with no or few blocking/scattering elements in their surroundings, MS antenna heights are usually smaller than those of local, natural and man-made features. Typical values range from 1.5 or so for handheld terminals to 3 m for vehicular terminals. For other radiocommunication systems for TV broadcasting or fixed wireless access operating in the same frequency bands, the propagation channel will present a milder behavior given that, in these cases, the receive antennas are usually directive and are normally sited well above the ground. Both the shadowing effect on the direct signal and the amount of multipath is considerably reduced.

Other operating scenarios where both ends of the link are surrounded by obstacles are in indoor communications where walls, the ceiling or the various pieces of furniture will clearly determine the propagation conditions.

The frequencies used in mobile communications are normally above 30 MHz and the maximum link lengths do not exceed 25 to 30 km. Macrocells in current 2G (second generation, e.g., GSM) or 3G (third generation, e.g., UMTS) systems are much smaller. It must be taken into account that mobile communications are two-way and that the uplink (MS to BS) is power limited. This is especially so in the case of regular portable, handheld terminals. Furthermore, mobile system coverage ranges are short due to the screening effects of the terrain and buildings in urban areas. This makes frequency reuse possible at relatively short distances. This is also an important feature in mobile networks which require a great spectral efficiency for accommodating larger and larger numbers of users.

Currently 3G wireless systems are being deployed in the 2 GHz band while wireless LANs are beginning to be deployed in the 5 GHz band while, still, the 2.4 GHz band is the most popular for this application. Fixed access systems in licensed bands with ranges of several km to a few tens of km are being deployed in the 3.5 GHz band in Europe while in the Americas their assigned band is closer to 2 GHz. The 5 GHz band will also be used in unlicensed fixed access network applications. Very promising, short-range systems are being proposed at higher frequencies such as in the neighborhood of 60 GHz where gaseous absorption mechanisms dominate. Such phenomena are not dealt with in this book.

Two representative and extreme scenarios may be considered:

(a) the case where a strong direct signal is available together with a number of weaker multipath echoes, i.e., *line-of-sight* (LOS) conditions; and
(b) the case where a number of weak multipath echoes is received and no direct signal is available, *non line-of-sight* (NLOS) conditions.

Case (a) occurs in open areas or in very specific spots in city centers, in places such as crossroads or large squares with a good visibility of BS. Sometimes, there might not be a direct LOS signal but a strong specular reflection off a smooth surface such as that of a large building will give rise to similar conditions. This situation may be modeled by a Rice distribution for the variations of the received RF signal envelope: the *Rice case*. Under these conditions, the received signal will be strong and with moderate fluctuations (Figure 1.2). The Rice distribution is studied in Chapters 5 and 6.

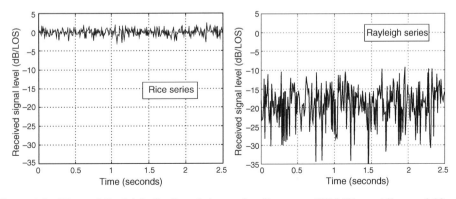

Figure 1.2   Rice and Rayleigh distributed time series. Frequency 900 MHz, mobile speed 10 m/s

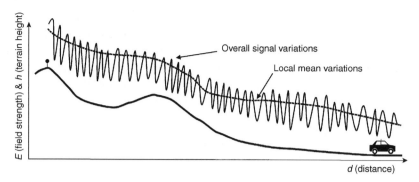

Figure 1.3   Variations in the received signal with the movement of the mobile [2]

Case (b) will typically be found in highly built-up urban environments. This is a worst-case scenario since the direct signal is completely blocked out and the overall received signal is only due to multipath, thus being weaker and subjected to marked variations (Figure 1.2). This kind of situation may also occur in rural environments where the signal is obstructed by dense masses of trees: wooded areas or tree alleys. The received signal amplitude variations in this situation are normally modeled with a Rayleigh distribution: the *Rayleigh case*.

The received field strength or the received voltage may be represented in the time domain, $r(t)$, or in the traveled distance domain, $r(x)$. Figure 1.3 shows a typical mobile communications scenario with MS driving away from BS along a radial route so that the link profile is the same as that of the route profile. The figure also shows a sketch of the received signal as a function of the distance from BS. The first thing to be noted is that the signal is subjected to strong oscillations as MS travels away from BS. The three rates of signal variation are also schematically represented.

For carrying out propagation channel measurements, the mobile speed, V, should remain constant. Of course, there are ways around this. In such cases, the traversed distance needs to be recorded too. In our simulations in later chapters and in the series analyzed in this chapter, we will assume a constant MS speed. For a constant V, it is straightforward to make the conversion between the representation in the time, $r(t)$, and traveled distance domains, $r(x)$ $(t = x/V)$.

Variable x may either be expressed in meters or in wavelengths. Based on such signal recordings plotted in the distance domain, it is possible to separate and study individually the fast and slow variations, due respectively to multipath and shadowing, as illustrated in Figure 1.4.

Generally, the received signal variations, $r(t)$ or $r(x)$, may be broken down, in a more or less artificial way, into two components [3]

- the slow or long-term variations: $m(t)$ or $m(x)$; and
- the fast or short-term variations: $r_0(t)$ or $r_0(x)$.

The received signal may, therefore, be described as the product of these two terms,

$$r(t) = m(t) \cdot r_0(t) \text{ or, alternatively, } r(x) = m(x) \cdot r_0(x) \quad (1.12)$$

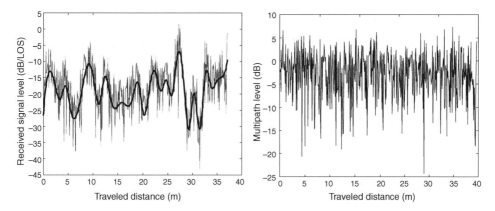

Figure 1.4   Overall and slow variations, and fast variations after removing the slow variations

when expressed in linear units. In dB, the products become additions, i.e.,

$$R(t) = M(t) + R_0(t) \quad \text{or, alternatively,} \quad R(x) = M(x) + R_0(x) \tag{1.13}$$

With this approach we are assuming that the fast variations are superposed on the slow variations. Figure 1.4 illustrates an overall time series where the slow variations are also plotted. The figure also shows the fast variations after removing (filtering out) the slow variations. The slow variations can be extracted from the overall variations through low-pass filtering by computing a running mean. This is equivalent to calculating the signal average for the samples within a route section of length $2L$ equal to some tens of wavelengths,

$$m(x_i) = \frac{\sum\limits_{k=-N}^{N} r_{i+k}}{2N + 1} \quad \text{for} \quad r_{i-N} \ldots r_i \ldots r_{i+N} \in x_i - L < x < x_i + L \tag{1.14}$$

Typically, lengths of 10λ to 40λ are used [3]. For example, for the 2 GHz ($\lambda = 0.15\,\text{m}$) band used in 3G mobile communications, the averaging length would be $2L \approx 3\text{–}6\,\text{m}$. The average value, $m(x_i)$, computed for a given route position x_i is usually called the *local mean* at x_i.

It has been observed experimentally [3] that the slow variations of the received signal, that is, the variations of the local mean, $m(x)$, follow a *lognormal distribution* (Chapter 6) when expressed in linear units (V, V/m, ...) or, alternatively, a *normal distribution* when expressed in logarithmic units, $M(x_i) = 20 \, \log m(x_i)$.

The length, $2L$, of route considered for the computation of the local mean, i.e., used to separate out the fast from the slow variations, is usually called a *small area* or *local area*. It is within a small area where the fast variations of the received signal are studied since they can be described there with well-known distributions (Rayleigh).

Over longer route sections ranging from 50 m or 100 m to even 1 or 2 km, the variations of the local mean are generally studied. This extended surface is usually called a *larger area*. Typically, standard propagation models do not attempt to predict the fast signal variations.

Instead they predict the mean, $\bar{M}(x) = \bar{E}(x)$, or, $\bar{M}(x) = \bar{V}(x)$, and the standard deviation (or *location variability*) σ_L of the local mean variations (normal distribution in dB) within the *larger area.*

Before low-pass filtering, the very slow variations due to the distance from BS must be removed. Free-space loss, L_{fs}(dB), is a very common model for the range-dependent loss. It is given by

$$l_{fs} = \frac{p_t}{p_r} = \left(\frac{4\pi d}{\lambda}\right)^2 \tag{1.15}$$

or, in practical units,

$$L_{fs}(dB) = 32.4 + 20\log f(MHz) + 20\log d(km) \tag{1.16}$$

where we have assumed isotropic antennas, i.e., with 0 dB gain or unit gain in linear units, with p_t and p_r the transmitted and received powers in W, and L_{fs} and l_{fs} the free-space loss in dB and in linear units (power ratio), respectively. Throughout this book, variables in capital letters will normally denote magnitudes expressed in logarithmic units (dB) and lower case letters will denote magnitudes expressed in linear units.

The free-space loss gives rise to a distance decay in the received power following an inverse power law of exponent $n = 2$ (Figure 1.5), i.e.,

$$p_1 \alpha \frac{1}{d_1^2} \quad \text{and} \quad p_2 \alpha \frac{1}{d_2^2} \quad \text{then, in dB,} \quad \Delta_p = 10\log\frac{p_2}{p_1} = 20\log\frac{d_1}{d_2} \tag{1.17}$$

where Δ_p is the received power difference in dB. The above expressions show a 20 dB/ decade (20 dB decrease when the distance is multiplied by 10) or 6 dB/octave (6 dB decrease when the distance is doubled) distance decay rate.

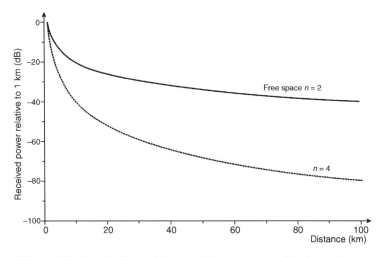

Figure 1.5   Received signal decay with distance: $n = 2$ and $n = 4$ laws

These variations are steeper in the first few kilometers of the radio path and become gentler for the longer distances. For example, using Equation 1.16, from km 1 to km 2, a 6 dB decrease takes place. However, the same 6 dB reduction is observed from km 10 to km 20.

It has been experimentally observed that, in typical mobile propagation paths, the signal's distance decay does not follow an $n = 2$ power law (as in free space) but, rather, it presents a larger exponent. Signal decay is usually modeled by an $l \alpha d^n$ law, i.e., l being proportional to the distance risen to the power n. The values of n typically are somewhere near 4, i.e., 40 dB/ decade (Figure 1.5). Widespread propagation models, briefly discussed in Chapter 3 such as those of Hata [4] and COST 231 [5], also predict exponents close to 4.

The path-loss expressions normally provided by propagation models (Chapter 3) are of the form

$$L(\text{dB}) = A + B \log d(\text{km}) = A + 10n \log d(\text{km}) \tag{1.18}$$

where A and n are dependent on the frequency and a number of other factors as listed below. Parameter A is the loss at a *reference distance*, in this case, 1 km, and n is the propagation decay law. Several factors, apart from the frequency and the distance that influence path loss, are taken into consideration by existing propagation models affecting the expressions for A and n. These factors are:

- the height of the MS antenna;
- the height of the BS relative to the surrounding terrain (*effective height*);
- the *terrain irregularity* (sometimes called *undulation*, Δh, or *roughness*, σ_t);
- the *land usage* in the surroundings of MS: urban, suburban, rural, open, etc.

When calculating the path loss, all such factors must be taken into account, i.e.,

$$L = L_{\text{Reference}} + L_{\text{Terrain Irregularity}} + L_{\text{Environment}} \tag{1.19}$$

The *path loss* is defined as that existing between isotropic antennas (0 dB gain). Isotropic antennas do not exist in practice but are commonly used in link budget calculations since they allow the definition of the loss independently of the antenna type. Then, when computing an actual link budget, the gains of the antennas to be used must be introduced in the calculations.

The *path loss* is made up of three main components: a *reference loss*, typically the free-space loss, although some models like Hata's [4] take the urban area loss as reference. Other models [6] use the so-called *plane-earth loss* ($n = 4$) as their reference (Chapter 8).

The second component is the loss due to *terrain irregularity* (we will be discussing this in some detail in Chapter 2) and, finally, the third component is the loss due to the *local clutter* or *local environment* where the additional loss will very much depend on the land usage in the vicinity of MS: urban, suburban, rural, open, woodland, etc.

1.3 Link Budgets

Here we briefly remind the reader of some of the basic elements in link budget calculations. In fact, link budgets are not only used for the wanted signal but also for noise and the interference. The equations below are valid for the wanted and the interference signals.

The *path loss* relates the transmit and received powers assuming no transmitter/receiver loss or gain (between isotropic antennas), i.e.,

$$L(\mathrm{dB}) = 10\log(p_t/p_r) \tag{1.20}$$

The path loss can be split into the sum of the *free-space loss* and the so-called *excess loss*,

$$L(\mathrm{dB}) = L_{\mathrm{fs}} + L_{\mathrm{excess}} \tag{1.21}$$

where the *excess loss* is given by,

$$L_{\mathrm{excess}}(\mathrm{dB}) = 20\log(e_0/e) \tag{1.22}$$

where e_0 is the field strength at the received antenna under free-space conditions and e is the actual field strength both in linear units (V/m).

So far we have assumed that the link does not contain gains or losses other than the path loss. If the gains and losses at both ends of the link are taken into account, the received power is given by

$$p_r = \frac{p_t g_t g_r}{l_t l l_r} \tag{1.23}$$

given in linear units or, in logarithmic units,

$$P_r(\mathrm{dBW\ or\ dBm}) = P_t + G_t + G_r - L_t - L - L_r \tag{1.24}$$

where l_t, or alternatively, L_t when using dB, is the loss at the transmit side, e.g., cables, etc., and g_t (G_t) is the transmit antenna gain.

A frequently used parameter for describing the radiated power is the EIRP (*equivalent isotropic radiated power*) defined as

$$\mathrm{eirp} = p_t g_t/l_t \tag{1.25}$$

in linear units or, in dB,

$$\mathrm{EIRP}(\mathrm{dBW\ or\ dBm}) = P_r + G_t - L_T \tag{1.26}$$

Link budgets are usually computed in terms of the received power. However, in some cases, the *field strength* (V/m) or the *power flux density* (W/m^2) are of interest, especially in interference studies. The received power can also be put in terms of the power flux density, ϕ, and the *effective antenna aperture*, A_e (m^2), i.e.,

$$p_r = \phi \cdot A_e \tag{1.27}$$

where the *effective antenna aperture* is related to the gain through

$$g_r = \frac{4\pi}{\lambda^2} A_e \tag{1.28}$$

where g_r is in linear units (power ratio). In dB, G_r (dB) $= 10 \log(g_r)$. The gain is frequently given with reference to an isotropic antenna and instead of dB, dBi units are often used. We will use both denominations indistinctly.

Coming back to the power flux density, it can be given in terms of the field strength at the receive antenna,

$$\phi = \frac{1}{2}\frac{|e|^2}{\eta} = \frac{1}{2}\frac{|e|^2}{120\pi} \text{ W/m}^2 \qquad (1.29)$$

where peak field strength values are used. If rms values were used the 1/2 in the right-hand side would be dropped. The term η is the impedance of free space which is equal to 120π. The power density at the receive antenna can also be put in terms of the transmitted power or the EIRP,

$$\phi = \frac{p_t g_t/l_t}{4\pi d^2} = \frac{\text{eirp}}{4\pi d^2} \text{ W/m}^2 \qquad (1.30)$$

By equating the two expressions for ϕ, we can find the expression for the field strength at the receive antenna under free-space conditions,

$$\frac{\text{eirp}}{4\pi d^2} = \frac{1}{2}\frac{|e|^2}{120\pi} \rightarrow |e| = \frac{\sqrt{60\,\text{eirp}}}{d} \text{ (V/m)} \qquad (1.31)$$

If the rms field strength is wanted, we should replace $\sqrt{60}$ by $\sqrt{30}$.

Finally, the well-known Friis equation for free-space conditions linking the transmit and receive powers can be reached by developing further Equation 1.27, thus,

$$p_r = \phi A_e = \frac{p_t g_t/l_t}{4\pi d^2}\frac{\lambda^2}{4\pi}g_r/l_r \qquad (1.32)$$

then

$$\frac{p_r}{p_t} = \frac{g_t g_r}{l_t l_r}\left(\frac{\lambda}{4\pi d}\right)^2 = \frac{g_t g_r}{l_t l_r l_{fs}} \qquad (1.33)$$

where l_{fs} is the free-space loss given in Equation 1.15. In the above expressions we should include other terms such as those due to impedance or polarization mismatch when necessary.

Two other link budgets associated with two basic propagation mechanisms are also presented next: one corresponds to *specular reflections* and the other to *diffuse scattering* on small objects, or on non-specular objects or rough surfaces such as the terrain. Other propagation mechanisms such as diffraction are dealt with in some detail in Chapter 2.

When a surface such as the ground or a building face is large and smooth, specular reflections can take place and reach the receiver provided that Snell's law is fulfilled, i.e., the angle of incidence and the angle of reflection are equal. In this case, the received power is ruled by the formula

$$\frac{p_r}{p_t} = \left[\frac{\lambda |R|}{4\pi(d_1 + d_2)}\right]^2 \qquad (1.34)$$

where the magnitude of Fresnel's reflection coefficient for the relevant polarization is included. Later in Chapter 8, the expressions of the complex reflection coefficients for the vertical and horizontal polarizations are given. For a power budget, the phase of the reflection coefficient is not needed, only its absolute value squared. However, if several rays combine coherently at the receiver their phases must be accounted for.

When the scattering object is not flat and smooth, or it is small, it will not show the same properties of specular reflectors. In this case, the scattered energy is shed in all directions or, possibly, within a given angular sector. Here, the link budget for diffuse reflections on this type of obstacles is ruled by the *bistatic radar equation*,

$$\frac{p_r}{p_t} = \frac{g_t}{l_t}\frac{1}{4\pi d_1^2}\sigma\frac{1}{4\pi d_2^2}\frac{\lambda^2}{4\pi}\frac{g_r}{l_r} \qquad (1.35)$$

where the obstacle parameter is its *radar cross-section*, σ (m^2).

This propagation mechanism produces much smaller contributions than those of specular reflections even for very large values of σ. This is due to the fact that two squared distances appear in its link budget in lieu of one distance (more specifically, the sum of two distances) as in the specular reflection case. Throughout this book, when modeling multipath, we will be using a multiple point-scatterer model. The powers for their corresponding contributions should fulfill the bistatic radar equation.

So far we have discussed static power budgets. In fact, the received signals are time varying. To account for such variations, when computing the system outage probability or setting up fade margins, the distribution of the variability, fast and slow, of the received signal must be well known. A good statistical knowledge of the signal variability behavior is of paramount importance for an optimal system planning. In addition, there is also the variability in the various interference sources (Chapter 3) and the noise.

1.4 Projects

We have reviewed in a concise way some relevant issues in wireless propagation modeling, especially those related to path loss and signal variability. Other concepts will be presented later in the book but, for the time being, this should be enough to get started. Now, we want to address some basic analysis techniques for time series, either measured or synthesized. We will do this on a step-by-step basis by introducing new concepts with each new project.

Project 1.1: Fast Fading Series

In this project, series11.mat (Figure 1.6) is supplied for analysis. It corresponds to a short section of simulated signal (in dBm) assumed to be received under homogeneous multipath conditions. This series, even though simulated, could just as well be a measured one.

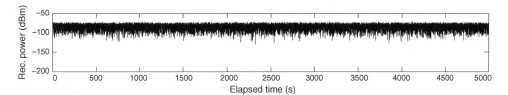

Figure 1.6 `series11` processed in Project 1.1

One possible way of recording a measured series could be in dB units, e.g., dBm (dB relative to 1 mW) as is the case here. This could correspond to a recording with a spectrum analyzer or field strength meter. In other cases, the measured series could be given in terms of analog to digital converter (ADC) units which must be translated into voltage or power units. In other cases, as in Project 1.3, the series can be given in terms of its in-phase and quadrature components. File `series11.mat` contains a two-column matrix which includes, in the first column, the time axis in seconds and, in the second column, the power in dBm.

Signal variations caused by multipath, in the case where the direct signal is assumed to be totally blocked, are usually represented by a Rayleigh distribution when expressed in units of voltage. We will see later in Chapter 5 how, if the magnitude were the power, we would have an *exponential distribution*. The *probability density function*, pdf, of the Rayleigh distribution is given by

$$f(r) = \frac{r}{\sigma^2} \exp\left(-\frac{r^2}{2\sigma^2}\right) \quad \text{for } r \geq 0 \tag{1.36}$$

where r is a voltage which actually represents, $|r|$, the magnitude of the complex envelope. We have dropped the magnitude operator for simplicity. This distribution has a single parameter, its *mode* or *modal value*, σ. Other related parameters are given in Table 1.1 as a function of σ. Script `intro11` is provided, which was used for plotting the Rayleigh pdf in Figure 1.7.

By integrating the pdf, the *cumulative distribution function*, CDF, can be obtained, i.e.,

$$\text{CDF}(R) = \text{Prob}(r \leq R) = \int_0^R f(r)dr = 1 - \exp\left(-\frac{R^2}{2\sigma^2}\right) \tag{1.37}$$

Table 1.1 Rayleigh distribution parameters as a function of its mode

Mode	σ
Median	$\sigma\sqrt{2\ln 2} = 1.18\sigma$
Mean	$\sigma\sqrt{\pi/2} = 1.25\sigma$
RMS value	$\sigma\sqrt{2} = 1.41\sigma$
Standard deviation	$\sigma\sqrt{2 - \pi/2} = 0.655\sigma$

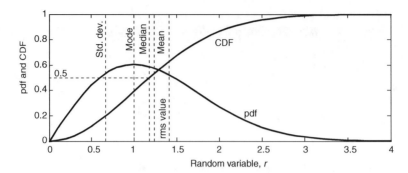

Figure 1.7 Probability density function and cumulative distribution function for a Rayleigh distribution with $\sigma = 1$. Generated with `intro11`

Rayleigh pdf and CDF functions are plotted in Figure 1.7 for $\sigma = 1$. The CDF is very useful when computing *outage probabilities* in *link budgets*. The CDF gives the probability that a given signal level is not exceeded. If this level is the system's *operation threshold*, this provides us with the probability that the signal level is equal or below such threshold, i.e., the *outage probability*. Knowing the CDF adequate *fade margins* can also be set up.

The parameters in Table 1.1 are defined as follows:

$$\text{mean}(r) = E[r] = \int_{-\infty}^{\infty} r\,f(r)dr = \sigma\sqrt{\frac{\pi}{2}} = 1.2533\sigma \tag{1.38}$$

$$\text{rms}^2(r) = E[r^2] = \int_{-\infty}^{\infty} r^2 f(r)dr = 2\sigma^2 \tag{1.39}$$

$$\text{variance}(r) = E[r^2] - E[r]^2 = 2\sigma^2 - \frac{\sigma^2\pi}{2} = \sigma^2\left(\frac{4-\pi}{2}\right) = 0.4292\sigma^2 \tag{1.40}$$

$$1 - \exp\left(-\frac{\tilde{r}^2}{2\sigma^2}\right) = 0.5, \ \text{thus, } \ \text{median}\,(r) = \tilde{r} = \sqrt{2\sigma^2\ln(2)} = 1.1774\sigma \tag{1.41}$$

where $E[\ ]$ is the expectation operator.

What we will do in this project (script `project11`) is analyze `series11` by computing its histogram (approximation of its pdf) and its sample CDF, and we will verify whether the series provided fits a Rayleigh distribution. The series is plotted in dBm in Figure 1.6. What we want is to model the voltage. A load resistance, R, of 50 Ω is assumed, thus, $p(\text{W}) = 1000 \times 10^{P(\text{dBm})/10}$. The power and the voltage are linked through $v = \sqrt{2Rp}$.

The resulting values of v are very small and awkward to handle. Hence, for working with voltages, we will normalize the series with respect to one of its parameters. The preferred option here is to keep using the expression for the pdf given above, i.e., as a function of the mode. However, the estimation of the mode from the time series is complicated. We can, however, estimate the mode from the mean in the case of a Rayleigh distribution by using the

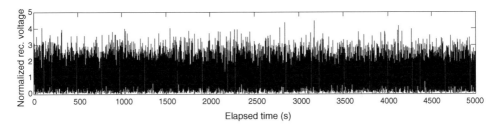

Figure 1.8   Voltage time series normalized with respect to its estimated modal value

equality $\sigma\sqrt{\pi/2} = 1.25\sigma$ in Table 1.1. Thus the series of v has been normalized with respect to its estimated modal value, σ, i.e., $v' = v/\sigma$ (Figure 1.8). The new series, v', should have a modal value equal to one. Another alternative would have been to use a version of the Rayleigh distribution put as a function of the mean. This would involve a very simple change of variable as shown below. The results would have been identical.

For completeness we show how the Rayleigh distribution can be expressed in terms of several of its parameters other than its mode. As a function of the mean the pdf and CDF are as follows,

$$f(r) = \frac{\pi r}{2\bar{r}^2}\exp\left(-\frac{\pi r^2}{4\bar{r}^2}\right) \quad \text{for} \quad r \geq 0 \quad \text{and} \quad P(R) = 1 - \exp\left(-\frac{\pi R^2}{4\bar{r}^2}\right) \tag{1.42}$$

where $\bar{r}$ is the *mean*. Put now as a function of the rms value, the pdf and CDF have the form,

$$f(r) = \frac{2r}{\overline{r^2}}\exp\left(-\frac{r^2}{\overline{r^2}}\right) \quad \text{for} \quad r \geq 0 \quad \text{and} \quad P(R) = 1 - \exp\left(-\frac{R^2}{\overline{r^2}}\right) \tag{1.43}$$

where $\sqrt{\overline{r^2}}$ is the *rms value*. Finally, as a function of the median,

$$f(r) = \frac{2r\ln(2)}{\tilde{r}^2}\exp\left(-\frac{r^2\ln(2)}{\tilde{r}^2}\right) \quad \text{for} \quad r \geq 0 \quad \text{and} \quad P(R) = 1 - \exp\left(-\frac{R^2\ln(2)}{\tilde{r}^2}\right) \tag{1.44}$$

where $\tilde{r}$ is the *median* of the distribution.

Now we go on to analyze the normalized series in Figure 1.8. Its average value is 1.25 which is equivalent to a modal value equal to one. We want to know whether the normalized series follows a Rayleigh distribution of mode one. In Figure 1.9 we plot the theoretical and the sample CDFs where we can observe that the match is reasonably good. The theoretical CDF has been computed using Equation 1.37 (`RayleighCDF`) and the sample CDF using function `fCDF`. Helpful MATLAB® (MATLAB® is a registered trademark of The Math-Works, Inc.) functions are `hist` and `cumsum`.

Figure 1.10 shows the theoretical and measured histograms where the theoretical values have been plotted using MATLAB® function `bar` and the equation

$$\text{Prob}(R_1 < r < R_2) = \int_{R_1}^{R_2} f(r)dr = \exp\left(-\frac{R_2^2}{2\sigma^2}\right) - \exp\left(-\frac{R_1^2}{2\sigma^2}\right) \tag{1.45}$$

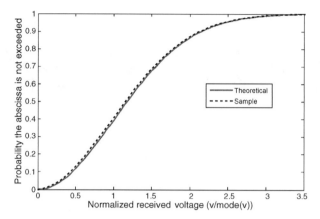

Figure 1.9   Time series and $\sigma = 1$ Rayleigh CDFs

where R_1 and R_2 are the bin limits (RayleighHIST). Again the match from a visual point of view is quite good. The histogram of the series can be computed using MATLAB® function hist that provides bin centers and associated frequencies, i.e., the number of occurrences. The frequencies have to be divided by the total number of samples in the series for converting them into probabilities.

Now, we are going to check whether the fit is good enough. As said, from a visual comparison between the measured and the theoretical CDFs and histograms, it is clear that the agreement is quite good. Now we want to *quantify how good the fit is*. This can be achieved by means of the *chi-square goodness-of-fit test* [7]. Other tests are also commonly used like the *Kolmogorov–Smirnov test* [7].

There are two basic elements in the chi-square test [7]. First, we must define *a measure* of the difference between the values observed experimentally and the values that would be expected if the proposed pdf were correct. Second, this measure has to be compared with *a*

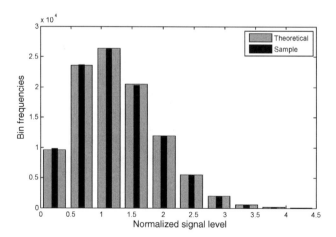

Figure 1.10   Time series and $\sigma = 1$ Rayleigh histograms

threshold which is determined as a function of the so-called *significance level* of the test. Usually this level is set to 1% or 5%. Below are the steps to be followed [7] for performing this test:

1. Partition the sample space, S_X, into the union of K disjoint intervals/bins.
2. Compute the probability, b_k, that an outcome falls in the k-th interval under the assumption that X has the proposed distribution. Thus, $m_k = nb_k$ is the expected number of outcomes that fall in the k-th interval in n repetitions of the experiment.
3. The chi-square measure, D^2, is defined as the weighted difference between the observed number of outcomes, N_k, that fall in the k-th interval, and the expected number, m_k,

$$D^2 = \sum_{k=1}^{K} \frac{(N_k - m_k)^2}{m_k} \qquad (1.46)$$

4. If the fit is good, then D^2 will be small. The hypothesis will be rejected if D^2 is too large, that is, if $D^2 \geq t_\alpha$, where t_α is the threshold for significance level α.

The chi-square test is based on the fact that for large n, the random variable D^2 has a distribution which approximately follows a chi-square with $K - 1$ degrees of freedom. Thus, the threshold, t_α, can be computed by finding the point at which (Figure 1.11) $\text{Prob}(X \geq t_\alpha) = \alpha$, where X is a chi-square random variable with $K - 1$ *degrees of freedom*, dof. The chi-square pdf is given by

$$f(x) = \frac{x^{(K-2)/2} e^{-x/2}}{2^{K/2} \Gamma(k/2)} \quad x > 0 \qquad (1.47)$$

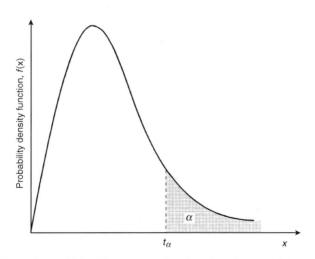

Figure 1.11   Threshold in chi-square test is selected so that $\text{Prob}(X \geq t_\alpha) = \alpha$ [7]

Table 1.2 Thresholds for significance levels 1% and 5%, and different degrees of freedom

K	5%	1%	K	5%	1%
1	3.84	6.63	12	21.03	26.22
2	5.99	9.21	13	22.36	27.69
3	7.81	11.35	14	23.69	29.14
4	9.49	13.28	15	25.00	30.58
5	11.07	15.09	16	26.30	32.00
6	12.59	16.81	17	27.59	33.41
7	14.07	18.48	18	28.87	34.81
8	15.51	20.09	19	30.14	36.19
9	16.92	21.67	20	31.41	37.57
10	18.31	23.21	25	37.65	44.31
11	19.68	24.76	30	43.77	50.89

where K is a positive integer and Γ is the Gamma function. This distribution is a special case of the Gamma distribution,

$$f_X(x) = \frac{\lambda(\lambda x)^{\alpha-1} e^{-\lambda x}}{\Gamma(\alpha)} \quad x > 0 \quad \text{and} \quad \alpha > 0, \lambda > 0 \tag{1.48}$$

when $\alpha = K/2$, K is a positive integer, and $\lambda = 1/2$.

The thresholds for the 1% and 5% levels of significance and different degrees of freedom are given in Table 1.2. The number of dof is $K - 1$, that is, the number of intervals or bins minus one. It is recommended that, if r is the number of parameters extracted from the data (e.g., mean, standard deviation, etc.), then D^2 is better approximated by a chi-square distribution with $K - r - 1$ degrees of freedom. Each estimated parameter decreases the degrees of freedom by one.

One further recommendation is on how the bins should be taken, since they may significantly influence the outcome of the test [7]. The selection of the intervals must be made so that they are equally probable. Another recommendation is that the expected number of outcomes in each interval be five or more. This will improve the accuracy of approximating the CDF of D^2 by a chi-square distribution.

We have performed the test first with the output of MATLAB® function `hist` which splits the range of observed values into intervals of equal length. In `project11` the number of bins has been set to 10. The resulting value of D^2 was 20.0873. This value has to be compared with that in Table 1.2 for $K = 10$ (intervals)$-1-1$ (the mode, parameter obtained from the sample) $= 8$. The value for 1% significance is 20.09 which barely exceeds the obtained value for D^2. In this case the test is passed. For 5% significance the threshold on the table is 15.51, meaning that the test is not passed.

As indicated above, it is convenient to use equally probable intervals. Thus, the test was carried out a second time with the intervals and frequencies given in Table 1.3 (`BINSequalprobRayleigh`). The final value of parameter D^2 is 15.9884, in this case the test for 5% significance is improved and the test is almost passed.

Table 1.3 Intervals and partial results of chi-square test with equal probability intervals

Bin Min.	Bin Max.	Measured frequency	Theoretical frequency	Elements of D^2
0	0.4590	10159	10000	2.5281
0.4590	0.6680	10179	10000	3.2041
0.6680	0.8446	10002	10000	0.0004
0.8446	1.0108	9782	10000	4.7524
1.0108	1.1774	9920	10000	0.6400
1.1774	1.3537	10152	10000	2.3104
1.3537	1.5518	9857	10000	2.0449
1.5518	1.7941	9970	10000	0.0900
1.7941	2.1460	9945	10000	0.3025
2.1460	∞	10034	10000	0.1156

If the MATLAB® version, including toolboxes, available to the reader contains function `gammainc`, the significance level can be computed by typing `alpha = 1-gammainc (0.5*chi2,0.5*df)`, where `chi2` is D^2 and `df` are the degrees of freedom.

Project 1.2: Shadowing Plus Multipath

For this project, file `series12.mat` is supplied. This series corresponds to a longer stretch of received signal where both shadowing and multipath effects are present. What we need to do is separate both variations in order to perform an independent study of the shadowing and the multipath-induced variations. Script `project12` is used in this analysis.

A section of the series in file `series12.mat` is shown in Figure 1.12. It corresponds to a signal at 2 GHz. File `series12.mat` contains a two-column matrix where the first column represents the traveled distance in meters and the second, the received signal in dBm. It is clear that both slow and fast variations are present. The separation is performed by means of a running mean filter implemented with a rectangular window that is slid through the series. MATLAB® function `conv` (convolution) has been used for this purpose. This process gives unreliable samples at the beginning and end of the filtered series which can be discarded.

The original series is in dBm and should be converted, as in Project 1.1, into voltage units. A window size of 10λ (variable `NofWavelengths`) has been used for separating the fast

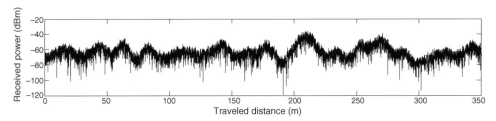

Figure 1.12 Original `series12` time series in dBm

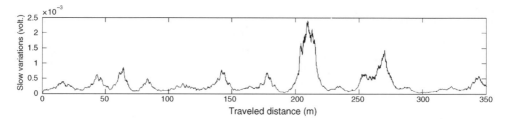

Figure 1.13 Slow variations in volts

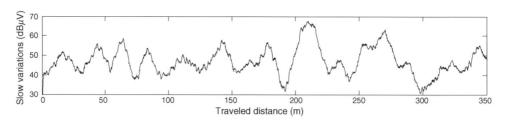

Figure 1.14 Slow variations in dBμV

and slow variations. Other window sizes can be tested. This is left for the reader to try. The sampling spacing is $\lambda/4$ (variable `SamplesperWavelength`). Figure 1.13 shows the running mean filtered voltage where the slow variations are clearly visible. High frequency components still remain but this is unavoidable. Figure 1.14 shows the filtered voltage in practical units, dBμV, dB relative to 1μV, i.e., $V(\text{dB}\mu\text{V}) = 20\log[v(\text{Volt}) \times 10^6]$.

We have first computed the local mean. By dividing the overall voltage by the local mean, we get a normalized voltage with respect to such local mean as shown in Figure 1.15. Figure 1.16 shows both the overall and the slow variations in dBμV. Note the effect of the convolution at the beginning of the plot.

A common assumption in the modeling of the fast variations is that they are Rayleigh distributed as already pointed out in Project 1.1. For the slow variations a lognormal, or normal for the variations in dB, is used. The lognormal distribution will be presented in Chapter 6. Here we remind the reader of some basic facts about the normal or Gaussian

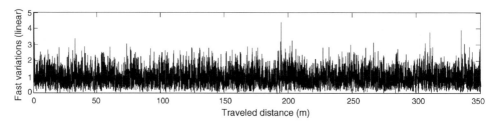

Figure 1.15 Fast variations in linear units

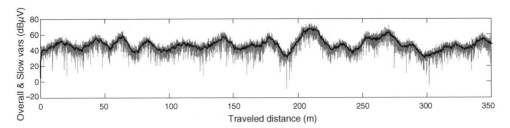

Figure 1.16   Slow and overall variations

distribution that we will need in this project. Its pdf is given by

$$p(x) = \frac{1}{\sigma\sqrt{2\pi}}\exp\left[-\frac{1}{2}\left(\frac{x-m}{\sigma}\right)^2\right] \tag{1.49}$$

and the cumulative distribution is

$$F(x) = \frac{1}{\sigma\sqrt{2\pi}}\int_{-\infty}^{x}\exp\left[-\frac{1}{2}\left(\frac{t-m}{\sigma}\right)^2\right]dt = \frac{1}{2}\left[1 + \operatorname{erf}\left(\frac{x-m}{\sigma\sqrt{2}}\right)\right] \tag{1.50}$$

where erf is the *error function*.

It is helpful to normalize the random variable x using its mean, m, and standard deviation, σ, i.e., $k = (x-m)/\sigma$, where k is the normalized Gaussian of zero mean and unity standard deviation. Another useful function is

$$Q(k) = \frac{1}{\sqrt{2\pi}}\int_{k}^{\infty}\exp(-\lambda^2/2)d\lambda \tag{1.51}$$

Function $Q(k)$ provides an easy way of calculating the probability that random variable x fulfills that $x > m + k\sigma$. This is equivalent to calculating the area under the tail of the pdf (Figure 1.17). Function Q is easily related to the error function or its complementary which

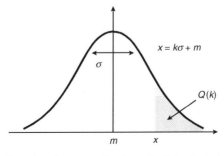

Figure 1.17   Function Q, area under the tail of the Gaussian distribution

Table 1.4 Relevant values of $Q(k)$

k	$Q(k) = 1 - F(k)$	k	$Q(k) = 1 - F(k)$
0	0.5	1.282	10^{-1}
1	0.1587	2.326	10^{-2}
2	0.02275	3.090	10^{-3}
3	1.350×10^{-3}	3.719	10^{-4}
4	3.167×10^{-5}	4.265	10^{-5}
5	2.867×10^{-7}	4.753	10^{-6}
6	9.866×10^{-10}	5.199	10^{-7}
		5.612	10^{-8}

are available in MATLAB® (functions `erf` and `erfc`) through

$$Q(k) = \frac{1}{2} - \frac{1}{2}\mathrm{erf}\left(\frac{k}{\sqrt{2}}\right) = \frac{1}{2}\mathrm{erfc}\left(\frac{k}{\sqrt{2}}\right) \tag{1.52}$$

Table 1.4 provides some practical values of the normalized Gaussian complementary CDF $(1 - F)$.

Coming back to our normalized, fast variations series, its mean is `1.0731` (see MATLAB® workspace) which is very close to the expected unitary mean. We will try and model the normalized voltage series by means of a Rayleigh distribution of mean 1.0731 which is equivalent to a Rayleigh with a modal value (Table 1.1) $\sigma = \mathrm{mean}/1.25 = 0.8585$.

Figure 1.18 shows a comparison between the sample and the theoretical CDFs. Figure 1.19 shows the CDFs of the slow and overall variations and in Figure 1.20 the sample CDF and a theoretical Gaussian CDF for the slow variations using the sample mean and sample standard deviation, `39.4916 dB` and `3.5708 dB` (MATLAB® workspace). Function `GaussianCDF` was also used in this study. The fittings are quite good as is to be expected since `series12` was produced using scripts from Chapter 6 implementing this lognormal–Rayleigh model (Suzuki model). In spite of the series provided being simulated, this should

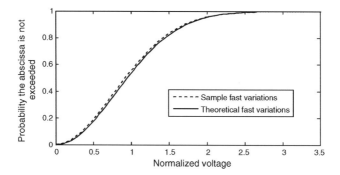

Figure 1.18 Sample and theoretical CDFs for the fast variations

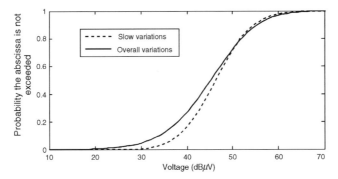

Figure 1.19   Sample CDFs for the slow and overall variations

be a good enough example of how real measurements are processed. Real measurements have been shown to respond quite well to this model [3].

Project 1.3: Complex Envelope Series

Files `series131.mat` and `series132.mat` contain simulated complex time series representing the in-phase, I, and quadrature, Q, parts of the received signal corresponding to a continuous wave (CW) transmission. The first corresponds to the Rayleigh case and the second to the Rice case. The supplied time series – `series131` and `series132` – are normalized with respect to the free-space signal level. The two files contain two-column matrices, the first storing the elapsed time, and the second, the complex envelope.

We are interested here in plotting some of the series parameters. For the Rayleigh case, script `project131` is used. Figure 1.21 shows the real or in-phase and imaginary or in-quadrature components as a function of time. They both should follow, as will be shown in Chapter 4, a Gaussian distribution. Figure 1.21 also shows an in-phase vs. quadrature plot where Lissajous-like curves can be observed which hint at some degree of periodicity in the signals: successive fades and enhancements. Figure 1.22 shows the magnitude of the signal, i.e., $|r| = |I + jQ|$ computed with MATLAB® function `abs` (absolute value). This series should

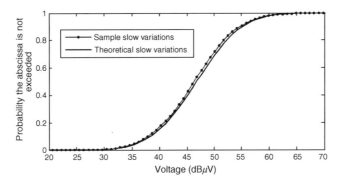

Figure 1.20   Sample and theoretical CDFs for the slow variations

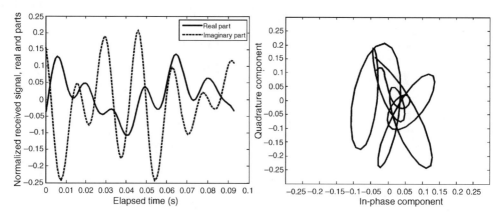

Figure 1.21   Rayleigh series. In-phase and quadrature components, and in-phase vs. quadrature

follow a Rayleigh distribution. Also in the same figure a plot of the magnitude in dB, i.e., $20 \log |I + jQ|$ is shown. Observe the deep fades and the large overall dynamic range of approximately 30 dB.

Figure 1.23 illustrates the phase (modulo-π) vs. time. This is obtained using MATLAB® function angle. The figure also shows the absolute phase variations obtained using MATLAB® function unwrap after angle. It is clear that the variations are totally random, i.e., neither increasing nor decreasing. The phase variations in this case are usually modeled with a *uniform distribution*.

For the Rice case, script project132 is used. Figure 1.24 shows the in-phase and quadrature components as a function of time. Again, they both should follow a Gaussian distribution. This figure also shows an *I* vs. *Q* plot. Now the Lissajous-like circles are traced about an in-phase level of one (direct signal). Figure 1.25 shows the magnitude of the signal computed with MATLAB® function abs (absolute value). This series should follow a Rice distribution which will be discussed later in Chapter 5. The figure also plots the magnitude in dB. Observe that now, the fades are not so deep, the dynamic range is much smaller and the

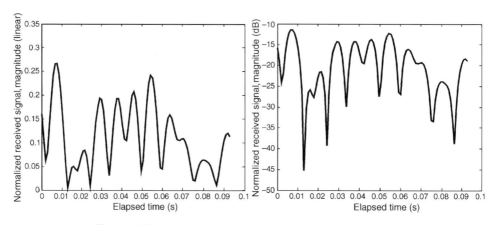

Figure 1.22   Rayleigh series. Magnitude in linear units and dB

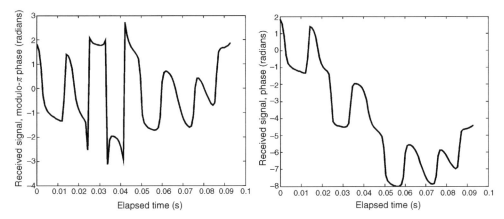

Figure 1.23   Rayleigh series. Modulo-π phase variations and absolute phase variations

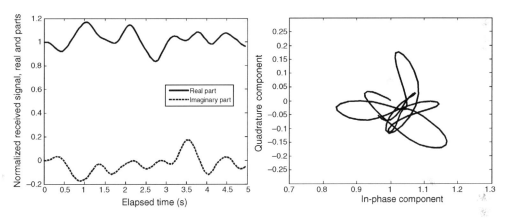

Figure 1.24   Rice series. In-phase and quadrature components, and in-phase vs. quadrature

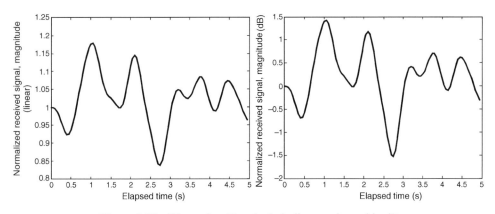

Figure 1.25   Rice series. Magnitude in linear units and in dB

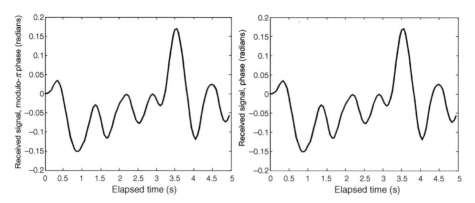

Figure 1.26   Rice series. Modulo-π phase variations and absolute phase variations

average level is higher, this due to the direct signal being much stronger than the multipath. Remember that, in the series, the normalized value of one (0 dB) represents the free-space level.

Figure 1.26 illustrates the phase (modulo-π) vs. time and the absolute phase variations. Unlike in the Rayleigh case, where the variations were totally random, here a clear trend, in this case about zero degrees, is observed in both plots. This clearly shows how the direct signal dominates over the multipath. We leave a more in-detail simulation and analysis of these phenomena for Chapters 4 and 5.

1.5 Summary

In this chapter, we have presented a brief introduction to the basic concepts and mechanisms driving the wireless propagation channel. Then, we presented very simple time-series analysis techniques which cover the basics previously introduced, these include the fast variations due to multipath and the combined effects of shadowing and multipath. Finally, we have looked into the complex envelope where we plotted its magnitude and phase. We have also presented the Rayleigh and Rice cases which correspond to harsh and benign propagation conditions. We will now go on to learn more about the various phenomena presented through simulation. This will allow us to become acquainted with a number of fairly complicated concepts in an intuitive way without the need to resort to involved mathematics. This will be done, as in this chapter, in a step-by-step fashion whereby new concepts will be presented as we progress in this familiarization process with the wireless propagation channel.

References

[1] L.W. Couch. Complex envelope representations for modulated signals (Chapter 1). In Suthan S. Suthersan (Ed.), *Mobile Communications Handbook*. CRC Press, 1999.
[2] J.M. Hernando & F. Pérez-Fontán. *An Introduction to Mobile Communications Engineering*. Artech House, 1999.
[3] W.C.Y. Lee. *Mobile Communications Design Fundamentals*. *Wiley Series in Telecommunications and Signal Processing*. John Wiley & Sons, Ltd, Chichester, UK, 1993.

[4] M. Hata. Empirical formula for propagation loss in land mobile radio services. *IEEE Trans. Veh. Tech.*, **29**(3), 1980, 317–325.

[5] Radiowave Propagation Effects on Next-Generation Fixed-Service Terrestrial Telecommunications Systems. Commission of European Communities. COST 235 Final Report, 1996.

[6] R. Edwards & J. Durkin. Computer prediction of service areas for VHF mobile radio network. *Proc. IEE*, **116**, 1969, 1493–1500.

[7] A. Leon-Garcia. *Probability and Random Processes for Electrical Engineering*, Second Edition (International Edition). Addison-Wesley, 1994.

Software Supplied

In this section, we provide a list of functions and scripts, developed in MATLAB®, implementing the various projects and theoretical introductions mentioned in this chapter. They are the following:

```
intro11          GaussianCDF
project11        Rayleighpdf
project12        RayleighCDF
project131       fCDF
project132       BINSequalprobRayleigh
                 RayleighHIST
```

Additionally, the following time series are supplied:

```
series11
series12
series131
series132
```

2

Shadowing Effects

2.1 Introduction

In Chapter 1 we presented the three different rates of signal variation as a function of the distance from BS: very slow, slow and fast. The range-dependent, very slow variations can be modeled as a power law, i.e., $1/d^n$, where the exponent, n, in typical cellular applications is close to 4. Superposed, we will observe slow variations due to shadowing and faster variations due to multipath.

Multipath effects will be analyzed later in Chapters 4 and 5. Typically, as we have seen in Chapter 1, slow and fast variations are studied separately. Their separation is performed by means of filtering. Once the faster variations have been filtered out what remains are the slower variations plus the range-dependent path loss.

The received power in logarithmic units, including the slow variations, can be represented by the expression

$$P(d) = P_{1km} - 10n \log(d_{km}) + X(0, \sigma_L) \tag{2.1}$$

where $P(d)$ is the received power at a distance $d(km)$, P_{1km} is the received power at 1 km, n is the power decay law and X is a Gaussian random variable of zero mean and standard deviation, σ_L, called the *location variability*. It should be pointed out that parameter P_{1km} is used in macrocell scenarios. In other cases where radio paths are shorter, e.g., indoor or urban microcells, other reference distances such as 1 m or 100 m are used.

Propagation loss models (Chapter 3) try to predict the mean of $P(d), \overline{P(d)}$, using expressions of the type

$$\overline{P(d)} = P_{1km} - 10n \log(d_{km}) \tag{2.2}$$

or, in terms of the *path loss*,

$$L = L_{1km} + 10n \log(d_{km}) \text{ dB} \tag{2.3}$$

where the path loss is often called *basic propagation loss*, and is defined as the loss between isotropic antennas.

Modeling the Wireless Propagation Channel   F. Pérez Fontán and P. Mariño Espiñeira
© 2008 John Wiley & Sons, Ltd

Sometimes, propagation models also provide *location variability* values as a function of the local terrain irregularity or the type of land usage: urban, suburban, rural, etc. When evaluating model *prediction errors* it is difficult to separate the actual errors from the location variability. Formally, the comparison must be made between the average of the slow- or long-term variations in the measurements (larger area mean), and keeping in mind that the fast multipath variations have already been filtered out.

Coming back to Equation 2.1, what it basically says is that the slow signal variations are normally distributed about an average value, which is what propagation models try to predict. The explanation for having a *normal distribution* can be given in terms of the *central limit theorem*, as the overall attenuation suffered by the link results from the addition of numerous individual shadowing processes leading up to a *Gaussian distribution*.

In this chapter we concentrate on the effect of individual, isolated obstacles of simple shapes (totally absorbing screens), to give the reader a quantitative idea of the order of magnitude of the attenuation to be expected. Later, in Chapter 3, we come back to the statistical representation of shadowing as a normally distributed process, when expressed in dB, which is superposed on an even slower decay rate proportional to the inverse of the radio path distance risen to the power of n.

Hence, here we will concentrate on the effect of absorbing screens (and associated apertures). One basic explanation for the non-zero received energy measured in the shadow of an obstacle is given by the Huygens–Fresnel principle [1]. This states that each point on a wavefront is itself a new source of radiation (Figure 2.1).

We will try to perform some simple calculations based on the Huygens–Fresnel principle. Here we start off with the scalar formulation due to Kirchhoff that implements the Huygens–Fresnel principle [2]. This scalar formulation neglects the effects of polarization, but provides reasonably good quantitative results for the purpose of this book. We assume [2] a closed screen, S, that surrounds a radiation source P_1 (Figure 2.2). This screen has an aperture, S_0, cut in it so that radiation exits the screen enclosed area. We are interested in determining the field strength reaching an observation point P.

The following expression [2] gives the electric field at P,

$$E(P) = \frac{j}{2\lambda} \int\!\!\!\int_{S_0} E_i(Q') \frac{\exp(-jkr)}{r} [\cos(\theta') + \cos(\theta)] dS' \tag{2.4}$$

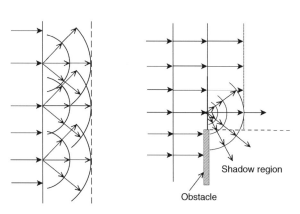

Shadow region

Obstacle

Figure 2.1 Huygens–Fresnel principle [1]

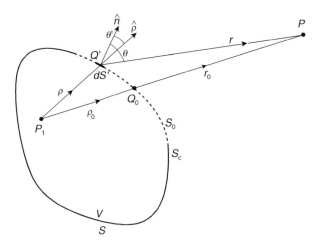

Figure 2.2   Assumed geometry in Kirchhoff's formulation [2]

where S_0 is the aperture surface, r is the distance from the aperture to P. Angles θ and θ' are graphically explained in Figure 2.2. Note that, unlike in the rest of this book, and for keeping consistency with the formulations in the references, we use capital letters for representing field strengths in linear units. Thus, $E_i(Q')$ is the incident field reaching the aperture from the source, given by

$$E_i(Q') = K \frac{\exp(-jk\rho)}{\rho} \tag{2.5}$$

where K is a constant depending on the transmitted power and antenna gain, and ρ the distance from the source to the screen. Introducing the expression of the incident field on that of the field at the observation point yields,

$$E(P) = \frac{j}{2\lambda} K \iint\limits_{S_0} \frac{\exp[-jk(\rho+r)]}{\rho r} [\cos(\theta') + \cos(\theta)] dS' \tag{2.6}$$

where a so-called *inclination factor* can be introduced, defined as

$$I(\theta, \theta') = \frac{1}{2} [\cos(\theta') + \cos(\theta)] \tag{2.7}$$

This factor quantifies the directivity of the aperture contributions as a function of the angular separation from the direct path. This term leads to a cardioid pattern, i.e.,

$$I(\theta) = I(\theta, \theta' = 0) = \frac{1}{2} [1 + \cos(\theta)] \tag{2.8}$$

when the aperture is *a sphere centered around the source*. This pattern shows very small directivity and is often neglected. The pattern has a maximum in the forward direction that gently decays to become zero in the backward direction. Including this factor, the field strength at the observation point is

$$E(P) = \frac{\mathrm{j}}{\lambda} \int\!\!\!\int_{S_0} E_i(Q') \frac{\exp(-\mathrm{j}kr)}{r} I(\theta, \theta') dS' \qquad (2.9)$$

and thus,

$$E(P) = \frac{\mathrm{j}}{\lambda} K \int\!\!\!\int_{S_0} \frac{\exp[-\mathrm{j}k(\rho + r)]}{\rho r} I(\theta, \theta') dS' \qquad (2.10)$$

Now we consider an aperture on *an infinite plane screen*. This is a very practical configuration that will allow us to perform very useful calculations. Let us apply the Kirchhoff diffraction integral to this case, i.e., an infinite plane screen with an arbitrarily shaped aperture (Figure 2.3).

When distances d_1 and d_2 to a point at the aperture are much larger than its maximum size, the inclination factor $I(\theta) = I(\theta, \theta' = 0)$ varies very little over the aperture and we may take it out and put it in front of the integral. Also, if points P_1 and P are not far from the z-axis, we may expand ρ and r around d_1 and d_2 [2] (Figure 2.3).

In this case, the diffraction integral can be put as follows,

$$E(P) = \frac{\mathrm{j}}{\lambda F_e} E_0(P) I(\theta) \int\!\!\!\int_{S_0} \exp[-\mathrm{j}k f(x', y')] dx' dy' \qquad (2.11)$$

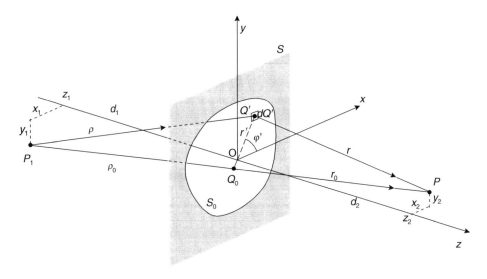

Figure 2.3 Arbitrary shaped aperture on an infinite plane screen [2]

where

$$E_0(P) \approx K \frac{\exp[-jk(d_1 + d_2)]}{d_1 + d_2} \qquad (2.12)$$

is the field strength at P of a free-space spherical wave radiated at P_1 and propagating along the straight path $P_1 P$, and

$$f(x', y') = \frac{1}{2F_e}[(x' - x_0)^2 + (y' - y_0)^2] \qquad (2.13)$$

where

$$F_e = \left(\frac{1}{d_1} + \frac{1}{d_2}\right)^{-1} = \frac{d_1 d_2}{d_1 + d_2} \qquad (2.14)$$

is the *equivalent focal length*. We are interested in evaluating the magnitude of the relative field strength, thus,

$$\frac{|E(P)|}{|E_0(P)|} = \left| \frac{j}{\lambda F_e} I(\theta) \int\int_{S_0} \exp[-jkf(x', y')]dx'dy' \right|$$

$$= \frac{1}{\lambda F_e} I(\theta) \left| \int\int_{S_0} \exp[-jkf(x', y')]dx'dy' \right| \qquad (2.15)$$

To avoid performing the above integrations numerically (as it will be done in Projects 2.1 and 2.6), we will introduce two mathematical functions (Fresnel's sine and cosine) that will help us solve simple but important diffraction problems. To achieve this, we introduce two new dimensionless variables u and v, which depend on the wavelength and the positions of the source point, P_1, the observation point, P, and an aperture point, Q'. The new variables are defined as follows,

$$u = \sqrt{2}\frac{x' - x_0}{R_1} \qquad \text{and} \qquad v = \sqrt{2}\frac{y' - y_0}{R_1} \qquad (2.16)$$

where x' and y' are generic points on the aperture, x_0 and y_0 are the coordinates of point Q_0, collinear with P_1 and P on the plane surface containing the aperture. Parameter

$$R_1 = \sqrt{\lambda F_e} = \sqrt{\lambda \frac{d_1 d_2}{d_1 + d_2}} \qquad (2.17)$$

is the *radius of the first Fresnel zone*. With these new variables, the aperture $S_0(x', y')$ is transformed into a new domain, $A_0(u, v)$, and the diffraction integral becomes

$$E(P) = \frac{j}{2}E_0(P) \int\int_{A_0} \exp\left[-j\frac{\pi}{2}(u^2 + v^2)\right]du\,dv \qquad (2.18)$$

where we have neglected the term $I(\theta)$. The diffraction equation can be put in the form

$$E(P) = E_0(P)F_d(u, v) \tag{2.19}$$

where

$$F_d(u, v) = \frac{j}{2} \iint_{A_0} \exp\left[-j\frac{\pi}{2}(u^2 + v^2)\right] du\, dv \tag{2.20}$$

is called the *Fresnel diffraction factor* [2] and is the inverse of the *excess loss* (Chapter 1) defined as

$$l_{\text{excess}} = |E_0|/|E|, \text{ or in dB}, L_{\text{excess}}(\text{dB}) = 10\log(l_{\text{excess}}) \tag{2.21}$$

i.e., the excess loss with respect to free space.

If we now consider a *rectangular aperture* (Figure 2.4) of dimensions

$$w_1 \le x' \le w_2 \quad \text{and} \quad h_1 \le y' \le h_2 \tag{2.22}$$

cut on an infinite screen [2]. The variables u and v become

$$u_1 = \sqrt{2}\frac{w_1 - x_0}{R_1}, \quad u_2 = \sqrt{2}\frac{w_2 - x_0}{R_1}, \quad v_1 = \sqrt{2}\frac{h_1 - y_0}{R_1} \quad \text{and} \quad v_2 = \sqrt{2}\frac{h_2 - y_0}{R_1} \tag{2.23}$$

The variables u and v are, in this case, mutually independent and the surface integral becomes the product of two separate linear integrals, i.e.,

$$F_d(u, v) = \frac{j}{2}\int_u \exp\left(-j\frac{\pi}{2}u^2\right) du \int_v \exp\left(-j\frac{\pi}{2}v^2\right) dv = F_d(u)F_d(v) \tag{2.24}$$

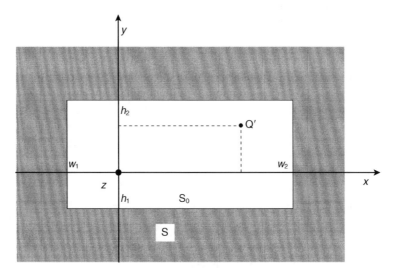

Figure 2.4 Rectangular aperture on an infinite screen [2]

The total diffraction factor becomes the product of the partial diffraction factors in u and v, i.e.,

$$F_d(u) = \sqrt{\frac{j}{2}} \int_{u_1}^{u_2} \exp\left(-j\frac{\pi}{2}u^2\right) du \quad \text{and} \quad F_d(v) = \sqrt{\frac{j}{2}} \int_{v_1}^{v_2} \exp\left(-j\frac{\pi}{2}v^2\right) dv \tag{2.25}$$

thus the total diffracted field is given by

$$E(P) = E_0(P)F_d(u)F_d(v) \tag{2.26}$$

The diffraction integrals can be calculated more easily in terms of the *Fresnel sine and cosine integrals* defined as follows,

$$\int_{u_1}^{u_2} \exp\left(-j\frac{\pi}{2}u^2\right) du = \int_0^{u_2} \exp\left(-j\frac{\pi}{2}u^2\right) du - \int_0^{u_1} \exp\left(-j\frac{\pi}{2}u^2\right) du$$

$$= \left[\int_0^{u_2} \cos\left(\frac{\pi}{2}u^2\right) du - j\int_0^{u_2} \sin\left(\frac{\pi}{2}u^2\right) du\right]$$

$$- \left[\int_0^{u_1} \cos\left(\frac{\pi}{2}u^2\right) du - j\int_0^{u_1=} \sin\left(\frac{\pi}{2}u^2\right) du\right] \tag{2.27}$$

$$= [C(u_2) - jS(u_2)] - [C(u_1) - jS(u_1)]$$

and

$$\int_{v_1}^{v_2} \exp\left(-j\frac{\pi}{2}v^2\right) dv = \int_0^{v_2} \exp\left(-j\frac{\pi}{2}v^2\right) dv - \int_0^{v_1} \exp\left(-j\frac{\pi}{2}v^2\right) dv$$

$$= \left[\int_0^{v_2} \cos\left(\frac{\pi}{2}v^2\right) dv - j\int_0^{v_2} \sin\left(\frac{\pi}{2}v^2\right) dv\right]$$

$$- \left[\int_0^{v_1} \cos\left(\frac{\pi}{2}v^2\right) dv - j\int_0^{v_1} \sin\left(\frac{\pi}{2}v^2\right) dv\right] \tag{2.28}$$

$$= [C(v_2) - jS(v_2)] - [C(v_1) - jS(v_1)]$$

where functions C and S are defined as

$$C(\tau_i) = \int_0^{\tau_i} \cos\left(\frac{\pi}{2}\tau^2\right) d\tau \quad \text{and} \quad S(\tau_i) = \int_0^{\tau_i} \sin\left(\frac{\pi}{2}\tau^2\right) d\tau \tag{2.29}$$

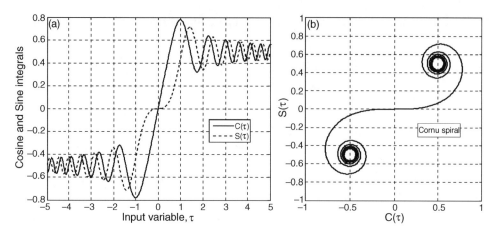

Figure 2.5 (a) Fresnel functions C and S, and (b) Cornu spiral: C vs. S generated with `intro21`

MATLAB® (MATLAB® is a registered trademark of The MathWorks, Inc.) provides these functions, `mfun('FresnelC',τ)` and `mfun('FresnelS',τ)`, and there is no need to program them. However, these functions belong to MATLAB®'s `symbolic toolbox` and might not be available to the reader. Hence, function `Fresnel_integrals` is provided that allows an alternative for calculating functions C and S.

Figure 2.5(a) illustrates C and S as a function of its input variable. Some interesting properties of C and S can be pointed out. Thus,

$$C(\tau) = -C(-\tau), \quad S(\tau) = -S(-\tau), \quad C(j\tau) = jC(\tau), \quad S(j\tau) = -jS(\tau) \qquad (2.30)$$

Relevant values of these functions are as follows,

$$C(\infty) = S(\infty) = \frac{1}{2}, \quad C(-\infty) = S(-\infty) = -\frac{1}{2}, \quad C(0) = S(0) = 0 \qquad (2.31)$$

Figure 2.5(b) shows the plot of C and S on a complex plane (Cornu spiral) where variable τ is equal the spiral's length from the origin. The spiral coils from point $-0.5 - j0.5(\tau = -\infty)$ to point $0.5 + j0.5(\tau = +\infty)$.

Coming back to the *rectangular aperture* case, the inverse of the excess loss is given by

$$
\begin{aligned}
F_d(u,v) &= F_d(u)F_d(v) \\
&= \sqrt{\frac{j}{2}}\{[C(u_2) - C(u_1)] - j[S(u_2) - S(u_1)]\}\sqrt{\frac{j}{2}}\{[C(v_2) - C(v_1)] - j[S(v_2) - S(v_1)]\}
\end{aligned}
$$
$$(2.32)$$

This formulation will be used again in Chapter 8 for quantifying the effect of a window in an outdoor-to-indoor link.

Now, we want to know what happens when the aperture is enlarged so that it becomes infinite, i.e., the screen totally disappears. The integration limits then go to infinite

(Figure 2.4), i.e.,

$$w_1 = -\infty \leq x' \leq w_2 = \infty \quad \text{and} \quad h_1 = -\infty \leq y' \leq h_2 = \infty \tag{2.33}$$

which leads to infinite values for u and v. Thus,

$$
\begin{aligned}
F_d(u,v) &= F_d(u)F_d(v)\\
&= \sqrt{\frac{j}{2}}\{[C(\infty) - C(-\infty)] - j[S(\infty) - S(-\infty)]\}\\
&\quad \times \sqrt{\frac{j}{2}}\{[C(\infty) - C(-\infty)] - j[S(\infty) - S(-\infty)]\}\\
&= \sqrt{\frac{j}{2}}[(0.5 + 0.5) - j(0.5 + 0.5)]\sqrt{\frac{j}{2}}[(0.5 + 0.5) - j(0.5 + 0.5)]\\
&= \sqrt{\frac{j}{2}}(1 - j)\sqrt{\frac{j}{2}}(1 - j) = 1 \cdot 1 = 1
\end{aligned}
\tag{2.34}
$$

This means that the excess loss is equal to one, that is, 0 dB, which means free-space conditions.

A very common model used in propagation studies is that of *knife-edge diffraction*. In many occasions, mountains or buildings are modeled as knife-edges for evaluating the excess loss they introduce. The knife-edge model assumes a semi-infinite aperture with limits (Figure 2.4)

$$w_1 = -\infty \leq x' \leq w_2 = \infty \quad \text{and} \quad h_1 \leq y' \leq h_2 = \infty \tag{2.35}$$

corresponding to limits in the u, v plane such that

$$u_1 = -\infty \leq x' \leq u_2 = \infty \quad \text{and} \quad v_1 \leq y' \leq v_2 = \infty \tag{2.36}$$

In this case, the surface integration becomes a one-dimensional integration

$$
\begin{aligned}
F_d(u,v) &= \frac{j}{2}(1-j)\int_{v_1}^{\infty}\exp\left(-j\frac{\pi}{2}v^2\right)dv\\
&= \frac{j}{2}(1-j)\left[\int_{0}^{\infty}\exp\left(-j\frac{\pi}{2}v^2\right)dv - \int_{0}^{v_1}\exp\left(-j\frac{\pi}{2}v^2\right)dv\right]\\
&= \frac{j}{2}(1-j)\{[C(\infty) - jS(\infty)] - [C(v_1) - jS(v_1)]\}\\
&= \frac{j}{2}(1-j)\{(0.5 - j0.5) - [C(v_1) - jS(v_1)]\}\\
&= \frac{j}{2}(1-j)\{[0.5 - C(v_1)] - j[0.5 - S(v_1)]\}
\end{aligned}
\tag{2.37}
$$

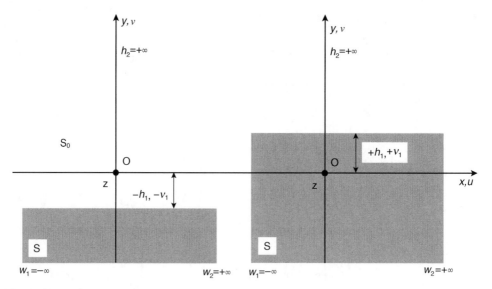

Figure 2.6 Knife-edge model. Two cases, on the left, negative obstruction or positive clearance. On the right, positive obstruction [2]

Figure 2.6 illustrates the geometry of the knife-edge diffraction model [2]. Figures 2.7 and 2.8 plot the results where two areas are defined: the *illuminated area* corresponding to negative values of v, and the *shadow area* corresponding to positive values of v. These results were obtained using script intro21.m.

It is quite common in LOS microwave radio link planning to define the so-called *clearance* parameter, c, which is the vertical distance from the center of the radio beam to the closest obstacle. Link clearance must be kept positive for most of the time even under sub-refractive conditions. In the case of the knife-edge model, we will be using the complementary of the clearance, i.e., the *obstruction* parameter, h, where $c = -h$. The

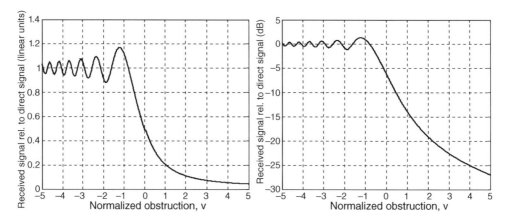

Figure 2.7 Knife-edge model. Relative field strength (in linear units and in dB) vs. v parameter

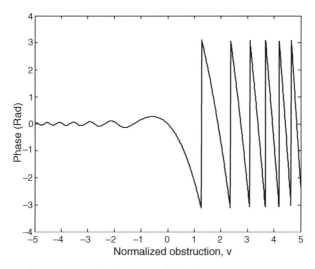

Figure 2.8 Knife-edge model. Phase vs. v parameter

obstruction will normally be normalized with respect to the radius of the first Fresnel zone, R_1. Parameter v is nothing more than the normalized obstruction multiplied by $\sqrt{2}$.

2.2 Projects

Now we go on to perform a number of simulations that will provide an idea of the magnitude of the shadowing attenuation caused by obstacles in the radio path.

Project 2.1: Knife-Edge Diffraction Calculations by Means of Numerical Integration

In this project we try to calculate the received field strength in the presence of an absorbing screen by *numerically integrating* the Fresnel–Kirchhoff equation. Reference [3] provides a procedure that develops the Fresnel–Kirchhoff diffraction equation for converting the problem into a two-dimensional one and where a number of simplifying assumptions (small wavelength, paraxial, etc.) were made. With reference to Figure 2.9, we present the propagation over a semi-infinite screen (knife-edge). The simplified expression provided in [3] is as follows,

$$E(x_{j+1}, z_{j+1}) = \sqrt{\frac{kx_j}{2\pi j x_{j+1} \Delta x_j}} \int_{\xi_j}^{\infty} E(x_j, z_j) \exp(-jkR) dz_j \qquad (2.38)$$

where $k = 2\pi/\lambda$, the lower integration limit, ξ_j, is the height of the knife-edge. $E(x_j, z_j)$ is the field strength at screen j. Distance R is computed on the $y = 0$ plane since the problem has been simplified to a two-dimensional one [3]. The distance can be calculated using

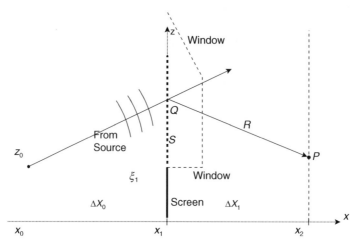

Figure 2.9 Schematic representation of the knife-edge model for numerically integrating the Fresnel–Kirchhoff equation [3]

the approximation

$$R = \Delta x_j + \frac{(\Delta z_j)^2}{2\Delta x_j} \tag{2.39}$$

We are interested in computing the excess loss, i.e., we want to compare the computed field strength at a given study point with that for free-space conditions at the same point, P. The expression for the free-space field strength at any point $(x, 0, z)$ is given by

$$E_{fs}(x, z) = \frac{C}{x} \exp\left[-jk\left(x + \frac{z^2}{2x}\right)\right] \tag{2.40}$$

where a spherical wave is assumed.

When performing the numerical integration of Equation 2.38, the *discretization step* must be selected to be sufficiently small (a fraction of the wavelength) so as not to introduce errors in the phase term. From a practical point of view, the *upper limit* of the integration cannot go all the way to infinity. This, in turn, truncates all alternating sign Fresnel zones from that point on, thus creating rippling on the calculated fields. To avoid this, we can multiply the field at the aperture, S, by a decaying window which gently falls to zero. We have used, for simplicity, a rectangular window followed by a section with a linearly decaying slope (Figure 2.9). Other windows such as Hamming, Hanning, etc. could be more effective.

In project21 we have calculated the relative field strength after a knife-edge. The frequency was 100 MHz, the transmitter was located at point $(-1000, 100)$ on the $x - z$ plane, the screen was located at $x = 0$ and the edge was at height of 100 m. The receiver was located at $x = 1000$ and an elevation scan from $z = 0$ to 250 m was performed. The aperture above the screen was sampled every $\lambda/8$. The rectangular part of the window went from

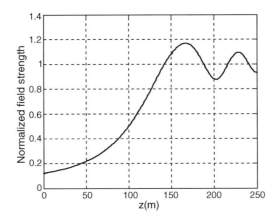

Figure 2.10 Computed normalized received field strength as a function of the z coordinate

$z = 100$ to 200 m afterward, the window falls off to zero at $z = 300$ m. Figure 2.10 illustrates the obtained result as a function of the z coordinate. It can be observed how the results match the theory.

Project 2.2: Knife-Edge Diffraction Calculations Using Functions C and S

In this project, we again simulate the effect of terrain irregularity by means of the knife-edge model. The geometry assumed is illustrated in Figure 2.11. A natural obstacle, e.g., a hill, is

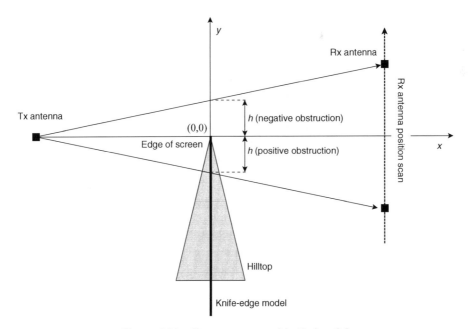

Figure 2.11 Geometry assumed in Project 2.2

between the transmitter and the receiver. This hill is modeled as a semi-infinite, completely absorbing screen. We want to observe the evolution of the received field strength as a function of the receive antenna height. We assume that the transmit antenna has the same height as the hilltop. In this case, we are going to use *Fresnel's sine and cosine functions* instead of performing a discrete integration.

We have to calculate the radio path height at the location of the hilltop, then compute the *obstruction*, h, or its inverse, the *clearance*, c. Then normalize it with respect to R_1 and multiply by $\sqrt{2}$ to compute parameter v_1. Parameter v_2 is infinite, and u_1 and u_2 are respectively minus and plus infinite. The settings in `project22` were as follows: frequency 2 GHz, location of transmitter `xt=-1000`, `yt=0`, location of screen top `xa=0` and `ya=0`, location of receiver `xr=1000` and `yr` goes from -50 to 50 with a step of 1 m.

The radio path is assumed to be a straight line between the transmitter and the receiver,

$$y = \left(\frac{y_r - y_t}{x_r - x_t}\right)(x - x_t) + y_t \tag{2.41}$$

The obstruction parameter, h, is then

$$h = y_a - \left[\left(\frac{y_r - y_t}{x_r - x_t}\right)(x_a - x_t) + y_t\right] \tag{2.42}$$

Parameter v_1 is given by

$$v_1 = \frac{\sqrt{2}h}{R_1} \tag{2.43}$$

where R_1 is the radius of the first Fresnel zone. From Section 2.1 and in MATLAB® code, the normalized field strength is given by

```
Enorm = (1-j)*j/2.*((0.5-mfun('FresnelC',v))-j*(0.5-mfun('FresnelS',v)))
```

Figure 2.12 illustrates the already well-known knife-edge diffraction plot where the relative field strength is presented in dB as a function of the antenna height. Again, two clear regions are observed: one for negative values of the antenna height that, in this case, as the transmitter is aligned with the edge, represent points in the *shadow area*; and the region, for positive values of the receiver antenna height, which represent the *lit area*.

In the shadow region, the received field gently falls off from the grazing incidence value of -6 dB to values lower than -20 dB. In the lit region, close to grazing incidence, the field increases from -6 dB to 0 dB, and then it oscillates around 0 dB due to the interference between the various unobstructed Fresnel zones.

Project 2.3: Simple Building Shadowing Model

In this project we try to use a slightly more complex, but still simple model for a building. In order to consider the limited width of a building and take into account the diffraction around

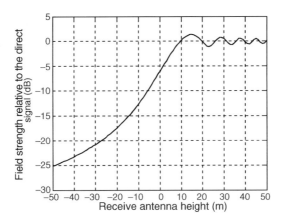

Figure 2.12   Relative field strength in dB as a function of receive antenna height

its sides, we will try to compute the two-dimensional integral in the diffraction model discussed in Section 2.1.

In this case, the screen is still infinitely wide but shows a protrusion representing the building. The mobile receiver is assumed to travel parallel to the edge of the screen which represents the street level (Figure 2.13). For calculating the double integral, this has to be split into three separate, contiguous surfaces $(\Sigma_1, \Sigma_2, \Sigma_3)$ to facilitate the calculation. The results of each individual integration have to be added coherently (magnitude and phase) for obtaining the overall normalized received field strength.

Figure 2.14 illustrates in more detail the geometric parameters needed for calculating u and v. These geometric parameters are given by the expressions,

$$\frac{y_0 - h_{Rx}}{dd} = \tan(\theta), \quad x_0 = D\tan(\varphi), \quad dd = \frac{D}{\cos(\varphi)} \tag{2.44}$$

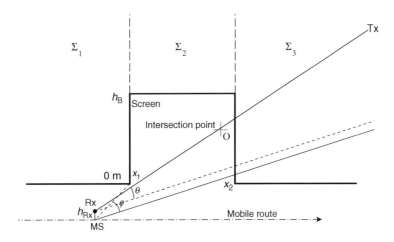

Figure 2.13   Model of a building as a protrusion on a knife-edge

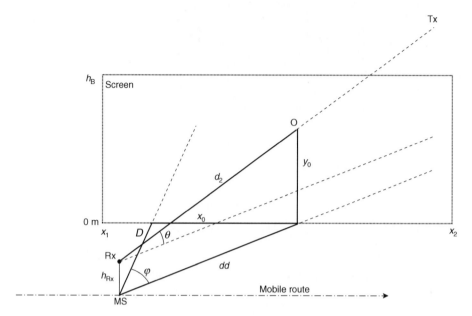

Figure 2.14   Detailed geometric parameters

and, thus,

$$y_0 = \frac{D \tan(\theta)}{\cos(\varphi)} + h_{Rx} \quad \text{and} \quad d_2 = \frac{y_0 - h_{Rx}}{\sin(\theta)} \tag{2.45}$$

For simplicity, we have also assumed that the distance d_1 from Tx to point O, intersection on the screen plane, is much larger than the distance d_2 from Rx to point O. In this case R_1 (*first Fresnel zone radius*) can be simplified to make it depend only on d_2, i.e.,

$$R_1 = \sqrt{\lambda \frac{d_1 d_2}{d_1 + d_2}} \approx \sqrt{\lambda d_2} \tag{2.46}$$

For the building part, the obstruction is given by $h_1 = h_B - y_0$. However, we need to define three sets of u and v parameters, one for each individual surface: Σ_1, Σ_2 and Σ_3. Thus, the following limits must be used for calculating the integrals,

$$u_{11} = -\infty, \quad u_{21} = \sqrt{2}\frac{x_1 - x_0}{R_1}, \quad v_{11} = -\sqrt{2}\frac{y_0}{R_1} \quad \text{and} \quad v_{21} = \infty \tag{2.47}$$

for surface Σ_1,

$$u_{12} = \sqrt{2}\frac{x_1 - x_0}{R_1}, \quad u_{22} = \sqrt{2}\frac{x_2 - x_0}{R_1}, \quad v_{12} = \sqrt{2}\frac{h_B - y_0}{R_1} \quad \text{and} \quad v_{22} = \infty \tag{2.48}$$

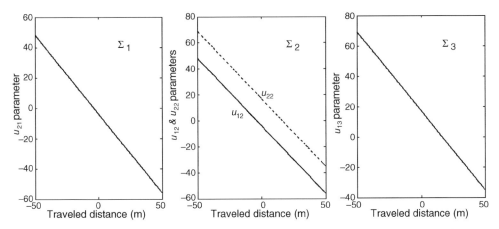

Figure 2.15 Evolution with the MS location of u_{21} for surface Σ_1, of u_{12} and u_{22} for surface Σ_2 and of u_{13} for surface Σ_3

for surface Σ_2, and

$$u_{13} = \sqrt{2}\frac{x_2 - x_0}{R_1}, \quad u_{23} = \infty, \quad v_{13} = -\sqrt{2}\frac{y_0}{R_1} \quad \text{and} \quad v_{23} = \infty \qquad (2.49)$$

for surface Σ_3.

The settings in `project23` are as follows (Figures 2.13 and 2.14): the frequency is 2 GHz, the Tx elevation is 30° and the azimuth 20°. The separation of MS from the building is 10 m, the MS antenna height is 1.5 m, the building height is 13 m and the coordinates for the beginning and end of the building are `x1=0` and `x2=20` respectively. The MS route is sampled every wavelength from coordinate `x=−50 to 50`.

Figure 2.15 illustrates the evolution of the u parameter as a function of the MS position for the three surfaces. Figure 2.16 illustrates the relative field strength along the MS route together with the outline of the building (only the most significant part of the route is shown). As Tx is on the right side of the building, its shade seems to be ahead of the actual building location.

In this project we have developed a very simple tool for modeling the shadowing effect of a single building. If we repeat this for several consecutive buildings, we can produce a simple street propagation simulator.

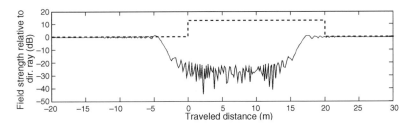

Figure 2.16 Relative field strength in dB and building outline

Project 2.4: Simple Street Model. Correlated Shadowing Series

In this project, we are interested in modeling the shadowing effects along a street. Basically, this means repeating the simulation carried out in the preceding project for several consecutive protrusions (representing individual buildings) on a semi-infinite screen. In addition to this, we want to introduce the concept of correlation or, rather, *cross-correlation* between the shadowing effects suffered by the signals from transmitters with different orientations and elevations with respect to the street. A more in-detail discussion on cross-correlation is presented later in Chapter 3.

We will be assuming the geometry shown in Figure 2.17. We suppose four buildings and three possible BS locations with different orientations with respect to the street. This scenario could represent a macro-diversity situation where a given street is served by three different BSs. Another hypothesis could be that of a wanted and two interfering links. We could be interested in reproducing what happens, at the MS, on several simultaneous links which are partially correlated. Later, in Chapter 3, we will generate partially correlated time series using a statistical approach. Here, we do it using a physical-geometrical approach.

The case studied in this project is slightly more complicated than that in Project 2.3, given that we have to take into account four buildings. Again we have to identify the integration limits for several non-overlapping, partial surfaces: $\Sigma_1, \Sigma_2, \Sigma_3, \Sigma_4, \Sigma_5, \Sigma_6, \Sigma_7, \Sigma_8$ and Σ_9, and do it for the three links. Thus, for link A,

$$u_{11a} = -\infty, \quad u_{21a} = \sqrt{2}\frac{x_1 - x_{0a}}{R_{1a}}, \quad v_{11a} = -\sqrt{2}\frac{y_{0a}}{R_{1a}} \quad \text{and} \quad v_{21a} = \infty \quad (2.50)$$

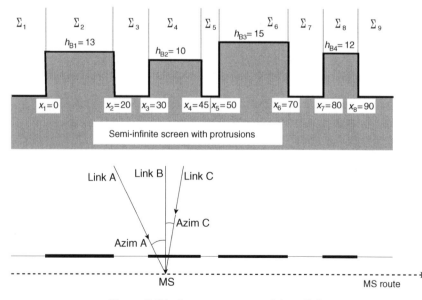

Figure 2.17   Street geometry and three links

for surface Σ_1,

$$u_{12a} = \sqrt{2}\frac{x_1 - x_{0a}}{R_{1a}}, \quad u_{22a} = \sqrt{2}\frac{x_2 - x_{0a}}{R_{1a}}, \quad v_{12a} = \sqrt{2}\frac{h_{B1} - y_{0a}}{R_{1a}} \quad \text{and} \quad v_{22a} = \infty \quad (2.51)$$

for surface Σ_2,

$$u_{13a} = \sqrt{2}\frac{x_2 - x_{0a}}{R_{1a}}, \quad u_{23a} = \sqrt{2}\frac{x_3 - x_{0a}}{R_{1a}}, \quad v_{13a} = -\sqrt{2}\frac{y_{0a}}{R_{1a}} \quad \text{and} \quad v_{23a} = \infty \quad (2.52)$$

for surface Σ_3,

$$u_{14a} = \sqrt{2}\frac{x_3 - x_{0a}}{R_{1a}}, \quad u_{24a} = \sqrt{2}\frac{x_4 - x_{0a}}{R_{1a}}, \quad v_{14a} = \sqrt{2}\frac{h_{B2} - y_{0a}}{R_{1a}} \quad \text{and} \quad v_{24a} = \infty \quad (2.53)$$

for surface Σ_4,

$$u_{15a} = \sqrt{2}\frac{x_4 - x_{0a}}{R_{1a}}, \quad u_{25a} = \sqrt{2}\frac{x_5 - x_{0a}}{R_{1a}}, \quad v_{15a} = -\sqrt{2}\frac{y_{0a}}{R_{1a}} \quad \text{and} \quad v_{25a} = \infty \quad (2.54)$$

for surface Σ_5,

$$u_{16a} = \sqrt{2}\frac{x_5 - x_{0a}}{R_{1a}}, \quad u_{26a} = \sqrt{2}\frac{x_6 - x_{0a}}{R_{1a}}, \quad v_{16a} = \sqrt{2}\frac{h_{B3} - y_{0a}}{R_{1a}} \quad \text{and} \quad v_{26a} = \infty \quad (2.55)$$

for surface Σ_6,

$$u_{17a} = \sqrt{2}\frac{x_6 - x_{0a}}{R_{1a}}, \quad u_{27a} = \sqrt{2}\frac{x_7 - x_{0a}}{R_{1a}}, \quad v_{17a} = -\sqrt{2}\frac{y_0}{R_{1a}} \quad \text{and} \quad v_{27a} = \infty \quad (2.56)$$

for surface Σ_7,

$$u_{18a} = \sqrt{2}\frac{x_7 - x_{0a}}{R_{1a}}, \quad u_{28a} = \sqrt{2}\frac{x_8 - x_{0a}}{R_{1a}}, \quad v_{18a} = \sqrt{2}\frac{h_{B4} - y_{0a}}{R_{1a}} \quad \text{and} \quad v_{28a} = \infty \quad (2.57)$$

for surface Σ_8, and

$$u_{19a} = \sqrt{2}\frac{x_8 - x_{0a}}{R_{1a}}, \quad u_{29a} = \infty, \quad v_{19a} = -\sqrt{2}\frac{y_{0a}}{R_{1a}} \quad \text{and} \quad v_{29a} = \infty \quad (2.58)$$

for surface Σ_9. We do not reproduce the rest of the limits for links B and C as they can be found within scrip `project24`, and can be looked up there.

The settings of simulator `project24` are as follows: the frequency was 2 GHz, all links were set to 30° elevation, link A has 10° azimuth, link B 0° azimuth and, finally, link C −30° azimuth. The separation of the MS route from the buildings is 10 m, and the MS antenna height was 1.5 m. The building dimensions are hB1=13, x1=0, x2=20, hB2=10, x3=30, x4=45, hB3=15, x5=50, x6=70, hB4=12, x7=80, x8=90. The route was sampled with a spacing of one wavelength.

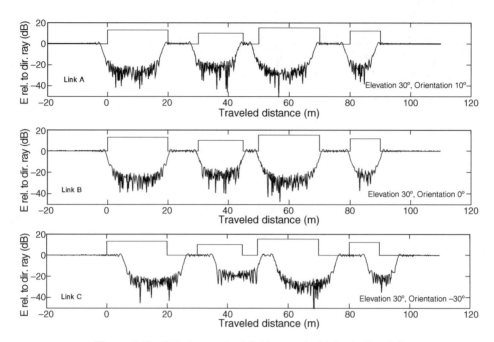

Figure 2.18   Relative received field strength. Links A, B and C

Figure 2.18 shows the relative received fields for links A, B and C in dB. Depending on the transmitter orientation, the series lag behind, or are in advance, with respect to the building locations, as can be observed in the figures. We have also calculated the cross-correlations between the magnitudes in linear units of the three relative field strengths. Before these calculations, their respective averages were removed. The cross-correlation is calculated in MATLAB® with function `xcorr`, where the parameter `coeff` was included. This function provides the cross-correlation for all possible relative lags between two signals. We are interested in simultaneous, i.e., zero-lag cross-correlations, this means we have to read the result at the center of the $2N - 1$ lag cross-correlation, where N is the length of the signal vectors. As seen in Figure 2.19, the maxima of the cross-correlation will not normally occur at the center point, thus reflecting the geometries of the links, i.e., the maxima and minima are displaced according to the orientation angle of the link with respect to the street. Another MATLAB® function, `corrcoef`, provides the zero-lag correlation coefficient.

Project 2.5: Terrain Effects. Four-Ray Model

In this project, we are going to explore further the knife-edge model by studying the case where reflections take place at both sides of the diffracting screen, as illustrated in Figure 2.20. In this case, we have to take into consideration that four sub-links are possible, the original Tx–Rx, plus others encompassing either or both the mirror images of Tx and Rx. Thus, the other links are: Tx′ (image of transmitter)–Rx, Tx–Rx′, Tx′–Rx′.

For calculating the relative field strengths in each sub-link, the same formulations as in `project22` were used in `project25`, taking good care of making the right geometric

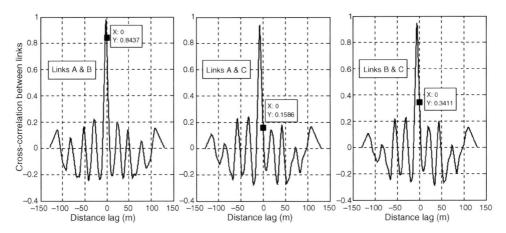

Figure 2.19 Cross-correlation between links A and B, A and C, and B and C

calculations for computing the *obstruction parameter* in each link. This means that the *obstruction* for the Tx–Rx link is given by

$$h = y_a - \left[\left(\frac{y_r - y_t}{x_r - x_t} \right) (x_a - x_t) + y_t \right] \tag{2.59}$$

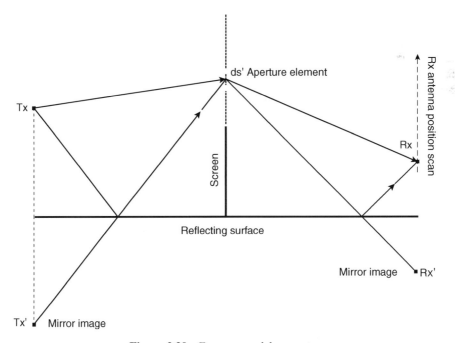

Figure 2.20 Four-ray model geometry

for the Tx′–Rx link,

$$h = y_a - \left[\left(\frac{y_r + y_t}{x_r - x_t} \right) (x_a - x_t) - y_t \right] \tag{2.60}$$

as now, the y-coordinate of the transmitter is $-y_t$. For the Tx–Rx′ link is

$$h = y_a - \left[\left(\frac{-y_r - y_t}{x_r - x_t} \right) (x_a - x_t) + y_t \right] \tag{2.61}$$

and, finally, for the Tx′–Rx′ link

$$h = y_a - \left[\left(\frac{-y_r + y_t}{x_r - x_t} \right) (x_a - x_t) - y_t \right] \tag{2.62}$$

After calculating the v parameters, the corresponding relative field strengths can be calculated using the expression

$$\frac{|E|}{|E_0|} = F_d(v) = \frac{j}{2} (1 - j) \{ [0.5 - C(v_1)] - j[0.5 - S(v_1)] \} \tag{2.63}$$

Before coherently adding the four contributions, it is necessary to add phase terms due to the different traveled paths, that is, in MATLAB® code for the Tx′–Rx contribution

```
d_DD=sqrt (((xr − xt)^2) + ((yt − yr).^2));
d_1R=sqrt (((xr − xt)^2) + ((yt + yr).^2));
Phase_Cor=exp(−j * kc * (d₁R − d_DD));
Enormalized2_p=Enormalized2.* Phase_Cor;
```

and the *reflection coefficients* at both sides of the screen, which we assumed to be equal to -1, given that we are close to gazing incidence. Figure 2.21 represents the overall and sub-link

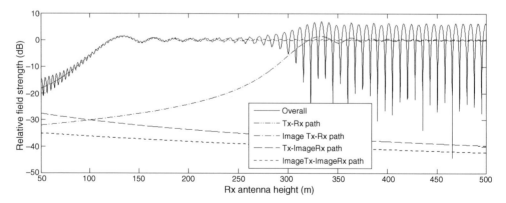

Figure 2.21   Overall relative received field strength in dB and individual contributions from the four sub-paths

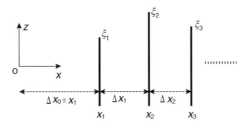

Figure 2.22   Multiple screen scenario

relative fields. Note the influence of the reflections on the strong enhancements and cancellations for the higher Rx antenna heights, and the rippling effects on the shadowed area.

Project 2.6: Other Screen Shadowing Scenarios

In this project we come back to the approach we took in Project 2.1, that is, we solve the simplified, 2D version of the Fresnel–Kirchhoff equation in [3] for several cases using a numerical integration approach. We reproduce, again, the formulas given earlier, i.e., the diffracted field at point (x_{j+1}, z_{j+1}) is given by

$$E(x_{j+1}, z_{j+1}) = \sqrt{\frac{kx_j}{2\pi j x_{j+1} \Delta x_j}} \int_{\xi_j}^{\infty} E(x_j, z_j) \exp(-jkR) dz_j \qquad (2.64)$$

where the distance in the phase term within the integral is $R = \Delta x_j + (\Delta z_j)^2/(2\Delta x_j)$.

Again, as in Project 2.1, we calculate normalized fields with respect to free space. Similarly, a window with a decaying characteristic has to be applied to the field at each aperture before propagating it on to the next one.

We tackled the multiple screen case in `project261` with reference to Figure 2.22 where several semi-screens are shown. The actual settings were as follows: the frequency was 2 GHz and four screens were simulated, the screen edges were at $z = 100$, the same as the transmitter. The spacings between screens and between the transmitter and the first screen, and the last receiver and the last screen was 1000 m. The actual simulated scenario is shown in Figure 2.23. Figure 2.24 shows the resulting normalized fields after each screen. In

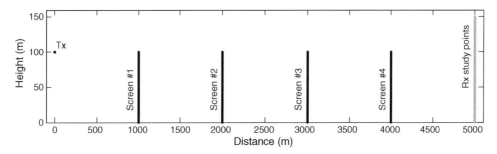

Figure 2.23   Multiple screen scenario in `project261`

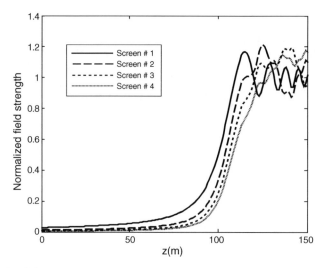

Figure 2.24 Normalized field strength behind screens #1, #2, #3 and #4

Chapter 3 we will briefly review a model for urban areas based on a multiple screen approach (Walfisch–Ikegami COST 235 model).

In `project262` we have simulated the case of a flat-topped obstacle, e.g., a building. The geometry of this case consists of two consecutive screens connected with a linear segment, as illustrated in Figure 2.25 [3]. For simplicity, we have used a reflection coefficient equal to -1. The actual simulated scenario is shown in Figure 2.26. Figure 2.27(a) illustrates the received signal on top of the second screen, due to the combined effect of the direct field from the first screen and the reflection on the horizontal segment joining the two screens. Figure 2.27(b) shows the normalized field after the flat-topped obstacle.

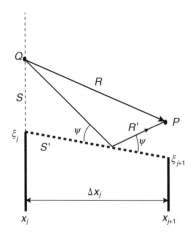

Figure 2.25 Reference geometry of where the linear segment joins two consecutive screens

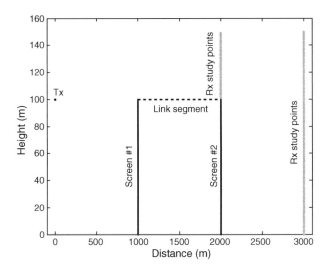

Figure 2.26   Actual simulated flat-topped obstacle

Lastly, in `project263`, we go one step forward and hint a possible way of analyzing the propagation loss over complex, irregular terrain described by linear piecewise profiles [3]. This is somehow the same as in the preceding simulation, where the reflected contributions have to be taken into account as well. Figure 2.28 illustrates a generic representation of the scenario to be simulated and Figure 2.29 shows the actual simulated terrain profile. Figures 2.30 and 2.31 show intermediate and final normalized field strengths. Note the strong attenuation after the second obstacle (wedge), and the interference pattern over the second obstacle, with strong enhancements and cancellations due to the effect of the reflections with coefficient −1.

The examples given are simple but should provide a good insight into the combined diffraction plus reflection effects. The reader is encouraged to analyze other irregular terrain configurations.

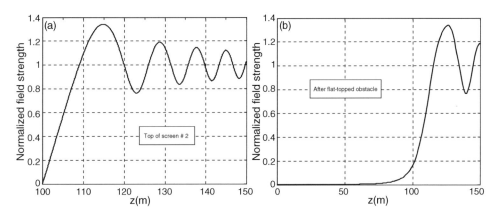

Figure 2.27   (a) Normalized field strength on top of second screen and (b) after flat-topped obstacle

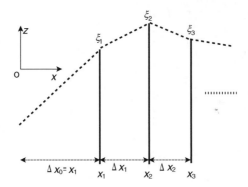

Figure 2.28   Irregular terrain modeled as a succession of screens joined by linear segments

Project 2.7: Terrain Effects. Edwards and Durkin Method

An alternative to the Fresnel–Kirchhoff equation for solving irregular terrain problems, is using simpler heuristic approaches based on approximations to the knife-edge diffraction curve. Here, we briefly discuss the Edwards and Durkin [4] model, and use it in `project27` for assessing the propagation loss or, rather, its inverse: the path gain, along a linear piece-wise profile.

For this model, a *reference loss* is defined which can either be the *free-space loss* or the *plane-earth loss* given by

$$L_{\text{pe}}(\text{dB}) = 120 - 20\log[h_{\text{t}}(\text{m})h_{\text{r}}(\text{m})] + 40\log d(\text{km}) \quad \text{or} \quad l_{\text{pe}} = \left(\frac{d^2}{h_{\text{t}}\,h_{\text{r}}}\right)^2 \tag{2.65}$$

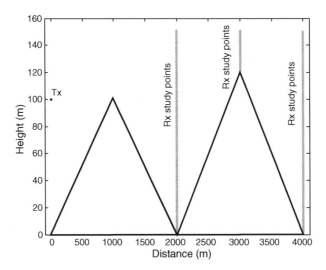

Figure 2.29   Simulated two-wedge scenario

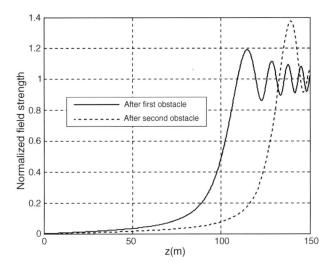

Figure 2.30 Normalized field strength after first wedge at a distance of 2000 m and after second wedge at a distance of 4000 m

which is discussed in some detail in Chapter 8. To compute the loss due to the terrain irregularity, several different situations or path types are defined, as listed in Table 2.1.

The calculation is performed for a detailed terrain profile obtained, for example, from a terrain database (TDB). The profile is then classified into one of the six path types listed in Table 2.1. Prior to carrying out this classification, the profile must be corrected to take into account the radius of the earth and the effect of refraction.

Normally, straight-line rays are used in propagation studies and it is assumed that the earth, instead of having a radius of $R_0 = 6370$ km, has an *effective radius*, kR_0, where factor

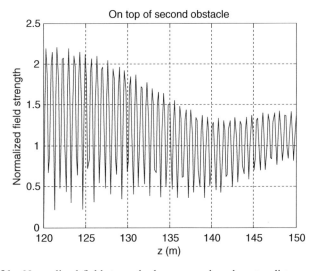

Figure 2.31 Normalized field strength above second wedge at a distance of 3000 m

Table 2.1 Types of radio paths considered in the Edwards and Durkin model [4]

| a | Line-of-sight | With sufficient path clearance |
b		With insufficient path clearance
c		1 obstacle
d		2 obstacles
e	Non line-of-sight	3 obstacles
f		More than 3 obstacles

k takes into account the refraction effects, especially for the longer distances. The usual value of k for a standard atmosphere is 4/3.

The profile is thus corrected from BS onward, with a parabola which approximates the earth's circumference, as shown in Figure 2.32. The parabola equation for correcting the terrain height samples is

$$y_i = h_i - \frac{x_i^2}{2k R_0}$$

(2.66)

where y_i is the corrected height, h_i is the height read on the TDB and x_i is the distance from BS. We will be calculating the path loss for all distances, x_i.

After correcting the profile, the next step is classifying it into one of the six categories in Table 2.1. For a direct line-of-sight path, a study of the clearance, c, must be carried out. Then, c must be compared with the radius of the first Fresnel zone, R_1, to verify whether the direct ray is clear from the most outstanding terrain feature by at least 60% of R_1. If this is the case, diffraction effects will be negligible.

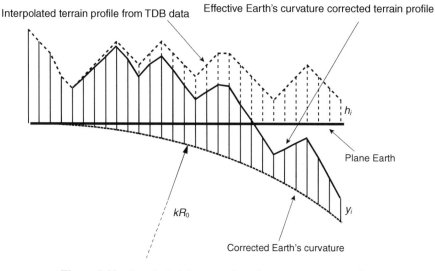

Figure 2.32 Terrain height corrections for area coverage studies

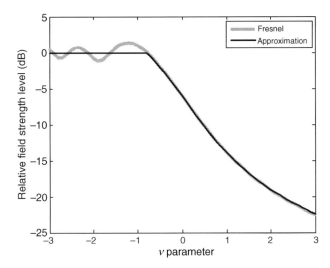

Figure 2.33 Diffraction losses as a function of the normalized obstruction parameter v. and approximation (`intro22`)

The diffraction loss, L_d, must be added to the reference loss, as indicated below. To compute L_d the curve shown in Figure 2.33 can be used or, alternatively, the following approximate formula [5],

$$L_d(v) = 6.9 + 20 \log \left[\sqrt{(v - 0.1)^2 + 1} + v - 0.1 \right] (\text{dB}) \qquad (2.67)$$

Figure 2.33, computed with `intro22`, shows the knife-edge diffraction curve together with its approximation which basically replaces the oscillating part, corresponding to line-of-sight conditions, with a constant zero dB loss.

Continuing with the line-of-sight path case, if it is observed that there is an insufficient clearance, i.e., $v > -0.8$ ($\approx 60\%$ R_1), the diffraction loss, L_d, must be computed. For case a *sufficient path clearance*, the total loss is equal to the free space loss, $L = L_{fs}$. For case b *insufficient path clearance*, the total loss will be the sum of the reference loss (maximum of free-space and plane-earth loss) and the loss due to diffraction, i.e., $L = \max(L_{fs}, L_{pe}) + L_d$. For case c a *single obstacle*, the total loss will follow the same expression, i.e., $L = \max(L_{fs}, L_{pe}) + L_d$.

For case d *two obstacles*, the total diffraction loss should be computed taking into account both obstacles. The procedure proposed is that due to Epstein and Peterson [6]. This method assumes that the link consists of two sub-links in tandem: the first between the transmitter and the second obstacle with diffraction on the first, and the second sub-link between the first obstacle and the receiver with diffraction on the second obstacle. Figure 2.34(a) schematically presents the procedure for calculating the obstructions h_1 and h_2. In this case, the total loss is given by $L = \max(L_{fs}, L_{pe}) + L_{d1} + L_{d2}$.

Case e, *three obstacles*, is illustrated in Figure 2.34(b). To compute the diffraction loss, the Epstein and Peterson method is applied successively to the three obstacles considering three sub-links as illustrated in the figure. The total loss is given by $L = \max(L_{fs}, L_{pe}) + L_{d1} + L_{d2} + L_{d3}$.

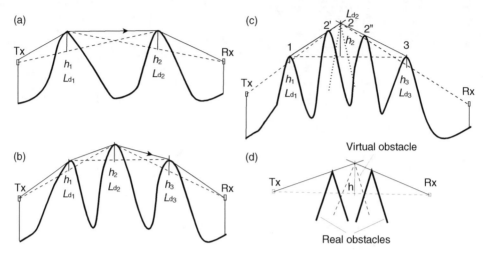

Figure 2.34 Epstein and Peterson method for (a) two and (b) three knife-edges. (c) JRC model for more than three knife-edges and (d) Bullington multi-edge diffraction method

Finally, case f, *more than three obstacles*, is shown in Figure 2.34(c). In this case, it is advisable to reduce the number of obstacles to a maximum of three, i.e., reduce the profile to case e. In order to do this, the procedure suggested by Bullington [7] is used. This involves describing the total diffraction effects by means of a *virtual obstacle* constructed as shown in Figure 2.34(d).

We have implemented, only in part, the above algorithm in `project27` for analyzing the terrain profile in Figure 2.35. Since the maximum path length is relatively small, the earth radius correction is not necessary. The reader is encouraged to verify this point. Figure 2.36

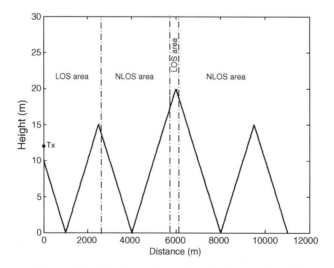

Figure 2.35 Terrain profile analyzed with `project27`

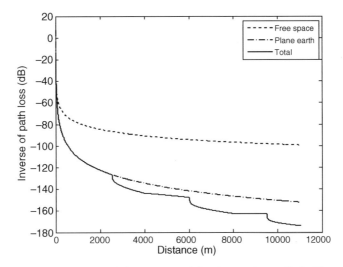

Figure 2.36 Inverse of path loss computed for the terrain profile in Figure 2.35

illustrates the inverse of the path loss, $-L$ (path gain), along the profile for a carrier frequency of 2 GHz, a transmit antenna height of 2 m and a receive antenna height of 1.5 m.

2.3 Summary

In this chapter we have presented some techniques for simulating the shadowing effects of screens. Even though the formulations presented are somehow more involved than those in other chapters, we have deemed it interesting to provide the reader with those tools to reproduce the effect of the presence of buildings and other large obstacles. In this way, for example, we have been able to reproduce the effects of the terrain or the signal variations while driving along a street. We also want to point out that through fairly simple physical models it is possible to reproduce the inherent cross-correlation properties observed between wanted or interfering signals converging on a given MS. Statistical alternatives are provided in the next chapter.

References

[1] J.D. Parsons. *The Mobile Radio Propagation Channel*, 2nd edition. John Wiley & Sons, Ltd, Chichester, UK, 2000.

[2] H.D. Hristov. *Fresnel Zones in Wireless Links, Zone Plate Lenses and Antennas*. Artech House, 2000.

[3] J.H. Whitteker. Fresnel–Kirchhoff theory applied to terrain diffraction problems. *Radio Sci.*, **25**(5), 1990, 837–851.

[4] R. Edwards & J. Durkin. Computer prediction of service areas for VHF mobile radio network. *Proc. IEE*, **116**, 1969, 1493–1500.

[5] ITU-R Recommendation P.526-9. Propagation by diffraction, 2007.

[6] J. Epstein & D.W. Peterson. An experimental study of wave propagation at 850 Mc. *Proc. IRE*, **41**(5), 1953, 595–661.

[7] K. Bullington. Radio propagation at frequencies above 30 Mc/s. *Proc. IRE*, October 1947, 1122–1136.

Software Supplied

In this section, we provide a list of functions and scripts, developed in MATLAB®, implementing the various projects and theoretical introductions mentioned in this chapter. They are the following:

```
intro21                      Fresnel_integrals
intro22                      erfz
project21                    epstein_peterson
project22                    triang_win
project23
project24
project25
project261
project262
project263
project27
```

3

Coverage and Interference

3.1 Introduction

In this chapter we will try and visualize the shadowing concept which gives rise to slow signal variations. As in the other chapters, in a first step, signal series will be synthesized. In this case, the series will be plotted as a function of the traveled distance.

We are going to review a number of issues related to shadowing effects affecting both the wanted signal and interference, and will visualize this shadowing effect together with the path loss. We have already reviewed in Chapter 1 how the transmitted signal, as it propagates away from the transmitter, experiences three different rates of variation, namely, very slow variations, also known as path loss, which are distance dependent, slow variations, mainly due to shadowing effects, and fast variations, due to multipath. Here we assume that the fast variations have been filtered out (as discussed in Chapter 1) and only path loss and shadowing variations remain.

When comparing the series produced in this way with a given *performance threshold*, either in a noise- or an interference-limited scenario, we will be assuming that the fast variations are taken into account by using modified thresholds, which have already been risen to account for the fast variations. Figure 3.1 illustrates this point [1].

The slow received signal variability due to shadowing, when expressed in dB, is usually assumed to follow a Gaussian distribution. When carrying out system planning, coverage quality is specified so that the operational threshold is exceeded for a given, large enough, percentage of locations (*L*%), e.g., 90%, 95%. Coverage is generally evaluated at the fringe of the service area. Overall *area coverage* quality is then obtained by extrapolating the *fringe coverage* quality to the whole of its inner area.

Figure 3.2(a) illustrates the concept of *fringe coverage* associated to a given probability level. This figure illustrates three contours corresponding to three probability levels: 50%, 90% and 95%. These probabilities mean that, if it were possible to drive along such contours, availability (i.e., exceedance of the threshold) percentages of 50%, 90% and 95% would be observed. Figures 3.2(b) and (c) schematically illustrate such test drives for the 50% and 90% probability levels [2].

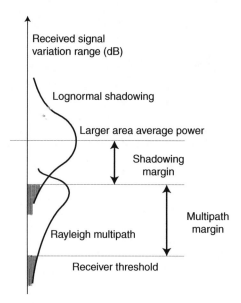

Figure 3.1   Slow and fast variations and associated margins [1]

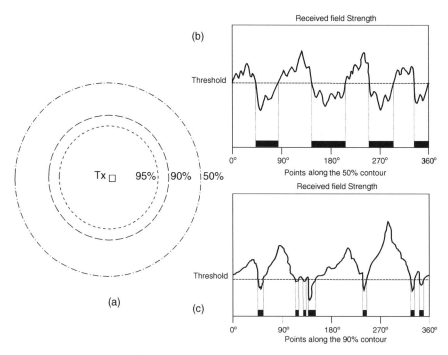

Figure 3.2   (a) Coverage contours for 50%, 90% and 95% of locations. Received signal levels while driving along the 50% (b) and 90% (c) contours [2]

If R is the radius of the coverage area (distance from BS to the cell's fringe), parameter F_u (area coverage probability) is defined as the fraction of the total area within the circle of radius R for which the signal level exceeds a given threshold, x_0. If $P(x_0)$ is the probability that the received signal x exceeds x_0 for an infinitesimal (incremental) area dA, then [3]

$$F_u = \frac{1}{\pi R^2} \int_{\text{Area}:\pi R^2} P(x_0)dA \quad \text{with} \quad P(x_0) = \text{Prob}(x \geq x_0) \quad (3.1)$$

Assuming that the mean received power, $\bar{x}$, is given by

$$\bar{x} = \alpha - 10\, n \log(r/R) \quad (3.2)$$

where, for convenience, the above expression has been slightly modified from that of a general propagation model given in Chapter 1, with α the average received power at a distance R. The slow signal variations due to shadowing follow a Gaussian distribution such that

$$p(x) = \frac{1}{\sigma_L \sqrt{2\pi}} \exp\left[-\frac{(x-\bar{x})^2}{2\sigma_L^2}\right] \quad (3.3)$$

The probability that x exceeds a given threshold x_0 at a distance $r = R$ (fringe) from BS, is (Chapter 1)

$$P(x_0, R) = \text{Prob}(x \geq x_0) = \int_{x_0}^{\infty} p(x)dx = \frac{1}{2} - \frac{1}{2}\text{erf}\left(\frac{x_0 - \bar{x}}{\sigma_L\sqrt{2}}\right) \quad (3.4)$$

For example [2], if the values of $\bar{x}$ and σ_L at a distance R are $-100\,\text{dBm}$ and $10\,\text{dB}$, respectively, and if the operation threshold is $x_0 = -110\,\text{dBm}$, the probability of exceeding the threshold level at a distance R is (Figure 3.3)

$$P(x_0, R) = \frac{1}{2} + \frac{1}{2}\text{erf}\left(\frac{1}{\sqrt{2}}\right) = 0.84 \equiv 84\% \quad (3.5)$$

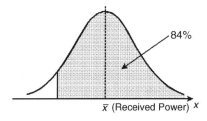

Figure 3.3 Area coverage probability in the example

Developing further the expression for $P(x_0, r)$, for any distance, r, from BS, we get

$$P(x_0, R) = \frac{1}{2} + \frac{1}{2}\operatorname{erf}\left[\frac{x_0 - \alpha + 10\,n\log(r/R)}{\sigma_L\sqrt{2}}\right] \tag{3.6}$$

The *area coverage probability*, F_u, can be calculated as follows,

$$F_u = \frac{1}{\pi R^2}\int_{\text{Area}:\pi R^2} P(x_0, r)dA \quad \text{with} \quad dA = r\,dr\,d\varphi \tag{3.7}$$

After some operations, the area coverage probability is given by [3][4],

$$F_u = \frac{1}{2}\left\{1 - \operatorname{erf}(a) + \exp\left(\frac{1 - 2ab}{b^2}\right)\left[1 - \operatorname{erf}\left(\frac{1 - ab}{b}\right)\right]\right\} \tag{3.8}$$

where $a = \dfrac{x_0 - \alpha}{\sigma_L\sqrt{2}}$ and $b = \dfrac{10\,n\log(e)}{\sigma_L\sqrt{2}}$.

In Figure 3.4, (intro31) curves for F_u (area coverage probability) as a function of the propagation parameters σ_L and n are shown. In the figure, the fractions of the area within a circle of radius R for which the received signal exceeds the threshold for several fringe probability values, $P(x_0, R)$, are given. For example, for $\sigma_L/n = 4$ and $P(x_0, R) = 0.5 = 50\%$ a value of $F_u \approx 0.7 = 70\%$ (*area coverage*) can be read. Typical location variability values, σ_L, range from about 5 to 10 dB.

Now that *terrain databases* (TDB) have became widely used for radio planning purposes, coverage calculations can be performed for small individual surface elements of, for example,

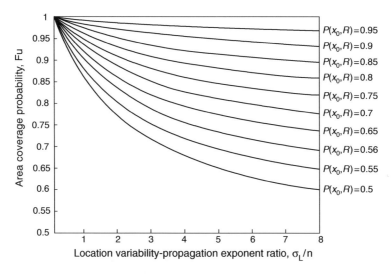

Figure 3.4 Relation between fringe coverage and area coverage probabilities for different propagation laws: n and σ_L

$100 \times 100\,\mathrm{m}^2$ or smaller. Prediction pixel sizes of $50 \times 50\,\mathrm{m}^2$, are quite common. If TDBs were used, the area of interest would be subdivided into such study areas for which the coverage probability, $P(x_0)$, could be computed. Later, all individual probabilities could be added up and averaged to compute the overall area coverage.

In order to provide an idea of how typical propagation models look like, we summarize next the main features and formulas corresponding to the well-known Hata model [5].

3.2 Hata Model

The Hata model [5] (intro32) is a version of the Okumura *et al.* model [6] developed for use in computerized coverage prediction tools. Hata obtained mathematical expressions by fitting the empirical curves provided by Okumura. Expressions for calculating the path loss, L (dB) (between isotropic antennas) for urban, suburban and rural environments are provided. For *flat urban areas*,

$$
\begin{aligned}
L(\mathrm{dB}) = &[69.55 + 26.16\log(f) - 13.82\log(h_\mathrm{t}) - a(h_\mathrm{m})] \\
&+ [44.9 - 6.55\log(h_\mathrm{t})]\log(d)
\end{aligned} \tag{3.9}
$$

where f is in MHz, h_t and h_m are in meters and d in km. Parameter h_t is the BS effective antenna height and h_m is the MS height, and d is the radio path length. For an MS antenna height of 1.5 m, $a(h_\mathrm{m}) = 0$. Model corrections are given next. Note the similarities between this model and the generic $L = A + 10n\log(d)$ model discussed in Chapter 1.

A number of corrections are given now for tailoring the model to more specific conditions. Thus, for a *medium-small city*,

$$
a(h_\mathrm{m}) = [1.1\log(f) - 0.7]h_\mathrm{m} - [1.56\log(f) - 0.8] \tag{3.10}
$$

For a *large city*,

$$
\begin{aligned}
a(h_\mathrm{m}) &= 9.29[\log(1.54 h_\mathrm{m})]^2 - 1.1 \quad f \leq 200\,\mathrm{MHz} \\
a(h_\mathrm{m}) &= 3.2[\log(11.75 h_\mathrm{m})]^2 - 4.97 \quad f \geq 400\,\mathrm{MHz}
\end{aligned} \tag{3.11}
$$

For a *suburban area*,

$$
L_\mathrm{s} = L - 2[\log(f/28)]^2 - 5.4 \tag{3.12}
$$

For *rural areas*,

$$
L_\mathrm{r} = L - 4.78[\log(f)]^2 + 18.33\log(f) - 40.94 \tag{3.13}
$$

The model is valid for the following range of input parameters: $150 \leq f(\mathrm{MHz}) \leq 1500$, $30 \leq h_\mathrm{t}(\mathrm{m}) \leq 200, 1 \leq h_\mathrm{m}(\mathrm{m}) \leq 10$ and $1 \leq d(\mathrm{km}) \leq 20$.

Some years ago, in view of the need to deploy higher frequency systems, such as the GSM at 1800 MHz or PCS at 1900 MHz, a new revision of the Hata model (COST 231-Hata [7])

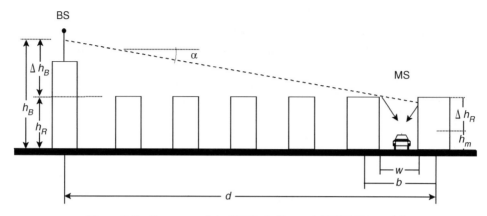

Figure 3.5   Geometry of the Walfisch–Ikegami COST 231 model

was developed using similar methods to those used by Hata. The COST 231-Hata model follows the expression

$$
\begin{aligned}
L = {} & [46.3 + 33.9 \log(f) - 13.82 \log(h_{\mathrm{t}}) - a(h_{\mathrm{m}})] \\
& + [44.9 - 6.55 \log(h_{\mathrm{t}})] \log(d) + C_{\mathrm{m}}
\end{aligned}
\tag{3.14}
$$

where $a(h_{\mathrm{m}})$ has the same expression as in the original model for a medium-small city and C_{m} is equal to $0\,\mathrm{dB}$ for medium-size cities and suburban centers, and equal to $3\,\mathrm{dB}$ for metropolitan centers. The validity of this modification is the same as for the original model, except for the frequency range which is $1500 \leq f(\mathrm{MHz}) \leq 2000$.

Due to the need for better characterizing the propagation in urban areas, within project COST 231 a model known as the Walfish–Ikegami model [7] was developed. The model's geometry is illustrated in Figure 3.5. The overall propagation loss is given as the sum of three terms,

$$
L = L_{\mathrm{fs}} + L_{\mathrm{rts}} + L_{\mathrm{msd}}
\tag{3.15}
$$

where L_{fs} is the free space loss, L_{rts} is the *rooftop-to-street* diffraction loss and L_{msd} is the *multi-screen diffraction loss* due to the buildings in between. This contribution is based on the Walfisch and Bertoni [8] model. In Chapter 2, we have briefly addressed the multiple edge diffraction case (`project261`).

3.3 Projects

Next we provide a series of simulation examples where we try to reproduce some of features relevant to the shadowing phenomenon affecting both the wanted signal and the interference.

Project 3.1: Generating Realistic Series with Path Loss and Shadowing

In this project we are going to first assume a small section of a circular traveled route, i.e., at a constant distance from BS. Such section of route could well represent a so-called *larger*

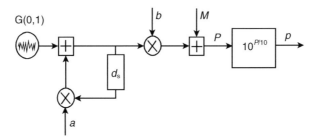

Figure 3.6 Low-pass filter method for generating slow signal variations due to shadowing

area (Chapter 1). We have, as inputs, the mean and standard deviation of the received power for this area. We will be generating a series of Gaussian distributed slow variations but forcing the autocorrelation properties present in reality. These can be characterized by the so-called *correlation distance*, L_{corr} (m).

The correlation distance characterizes the rate of change of the shadowing phenomenon. Typical L_{corr} values may be very different depending on the sizes of the obstacles (buildings, houses, trees, etc.) in the neighborhood of MS, going from a few meters to several tens of meters.

In `project311` we will be using an autoregressive filter [9] as illustrated in Figure 3.6. The input is a random number generator (`randn`) producing uncorrelated Gaussian samples of zero mean and unit standard deviation.

We set the sample spacing, d_s (m), and we want to synthesize a series with a correlation length, L_{corr} (m), with $L_{corr} > d_s$, and a location variability, σ_L dB. This is achieved using a filter with z-transform response

$$H(z) = \frac{1}{1 - az^{-1}} \tag{3.16}$$

which is a single pole $(z = a)$ IIR filter. Input samples are 'delayed' by d_s and multiplied by a and added to the current input sample, where $a = \exp(-d_s/L_{corr})$. Finally, the filter output is multiplied by $b = \sigma_L\sqrt{1 - a^2}$ to introduce the wanted location variability. Then the mean, M, is added. Finally, the series in dB can be converted to linear power units.

Figure 3.7 shows the simulated results. The following inputs were used: correlation length, `Lcorr=10` (m), sample spacing, `ds=1` (m), number of samples, `Nsamples=200`, and larger area mean and location variability `MM=−80` and `SS=5`. The figure illustrates the original uncorrelated Gaussian of unit standard deviation which has been shifted by `MM` for comparing it with the resulting series. Figure 3.7 also shows the autocorrelation functions of the original signal and the resulting one. For this calculation, MATLAB® (MATLAB® is a registered trademark of The MathWorks, Inc.) function `xcor` option `'coeff'` has been used. Note the delta-like autocorrelation for the series out of `randn` and the wider one for the filtered series.

Now, we go on to do the same simulation using a different approach. A straightforward way of producing correlated samples is by means of interpolation. `Project312` uses MATLAB® function `interp1` with option `'spline'`. Here, we use a random number generator with the wanted mean and standard deviation, and assume that the sample spacing is L_{corr}, as the samples generated by `randn` are uncorrelated. Through interpolation down to

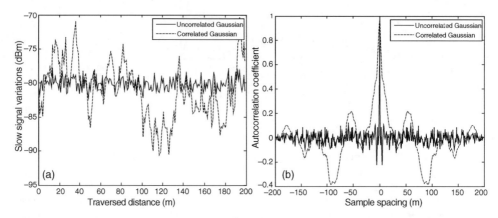

Figure 3.7 (a) Uncorrelated Gaussian series shifted by `MM=−80` and synthesized Gaussian series of mean `MM` and standard deviation `SS=5` with correlation length `Lcorr=10` m. (b) Autocorrelation functions of original series out of `randn` and filtered series

a sample spacing, d_s, we are able to introduce the needed autocorrelation distance in the generated series (Figure 3.8).

We performed simulations for a frequency of 2 GHz, i.e., a wavelength of 15 cm, a value that we also chose as our sample spacing, d_s. The correlation length is set to 30 m, the larger area mean was −70 dBm and location variability was 7 dB. Figure 3.9 illustrates the samples generated using `randn` (spaced L_{corr}) together with the corresponding interpolated values (spaced d_s).

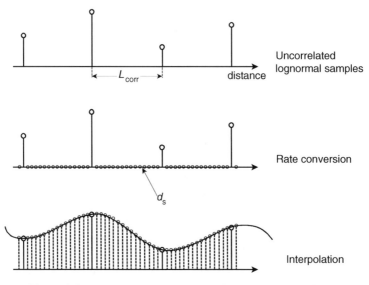

Figure 3.8 Generating correlated samples through interpolation

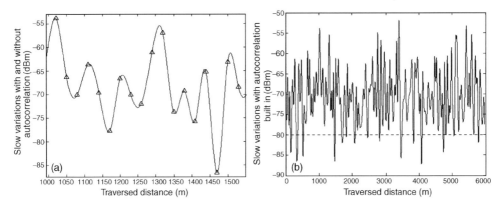

Figure 3.9 (a) Uncorrelated (triangles) and (b) correlated Gaussian signal variations, and fringe coverage probability calculations for a -80 dBm threshold

We are also interested in computing, both theoretically and from the synthesized series, what is the coverage probability. Figure 3.9 also illustrates a long series that could correspond to a circular route at a constant distance from BS. Also shown is the operation threshold set to -80 dBm. We have calculated directly from the series (using MATLAB® function find) the coverage probability, i.e., the probability of the received signal being above the threshold. The obtained value is CovProb=0.9231, i.e., 92.31%. Theoretically, the same probability can be calculated using function $Q(k)$ for $k = (-80 - (-70))/7 = -10/7 = -1.4286$, i.e., $Q(1.4286) = 0.9234$ (Chapter 1). The differences between theoretical values and simulations are usually due to the limited size of the simulated series.

In project313 we go one step further by introducing the *path loss* along a radial route several km long, starting at BS. Assuming a path loss given by a Hata-like (e.g., $A = 100$ and $n = 3.6$) model we calculate the average received power as a function of the distance from BS. This is the mean of the superposed shadowing variations which are further characterized by a location variability, σ_L dB, and a correlation length, L_{corr} (m).

The simulator settings are the same as in the previous project, where an EIRP $= 30$ dBm has been chosen. Figure 3.10 shows the resulting very slow and slow variations. The figure also illustrates a zoomed-in area showing the slow variations superposed on the path loss.

Project314 extends the preceding project to study a BS *handover* situation. MS is assumed to travel from BS1 to BS2 which are 6 km apart. Somewhere halfway between both BSs, the averages of the received signals are very similar in level. A *threshold* and a *margin* are defined so that two, very simple *HO algorithms* can be simulated. Figure 3.11 illustrates the two received signals and the *HO threshold* and *HO margin* above it. The first algorithm just performs the switch whenever the received signal is below the threshold and the alternative signal is larger. Figure 3.12(a) illustrates the numerous switches that take place. If a margin is taken into consideration in the HO algorithm so that, to switch back to the former BS, the difference between received signals must be above such margin, then the number of BS changes is drastically reduced (Figure 3.12(b)).

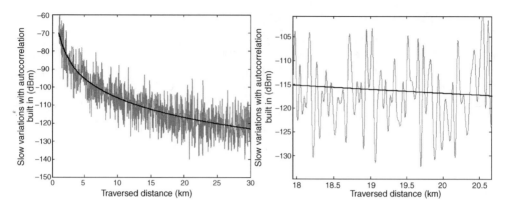

Figure 3.10 Simulated path loss plus slow variations due to shadowing, and zoomed-in section of the curve

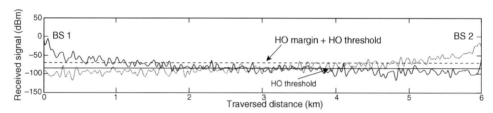

Figure 3.11 Signals from two BSs, and HO threshold and margin

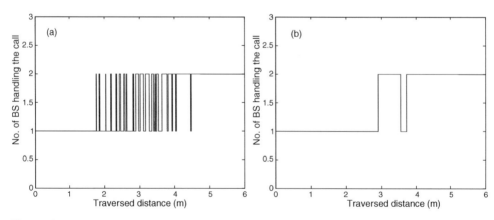

Figure 3.12 (a) BS handling the communication using a very simple, above-the-threshold HO algorithm. (b) BS handling the communications using a more complex HO algorithm using a margin

Of course, real HO algorithms are much more complex, and involve different criteria which are part of the network optimization process. Other, more complex algorithms are left for the reader to implement and try out.

Project 3.2: Generating Partially Cross-Correlated Shadowing Series

In this project, we go one step further in our analysis of the shadowing-induced variations, as we consider the cross-correlation between two signals, e.g., two wanted signals or a wanted and an interfering signal. In this case, in addition to showing the specific autocorrelation properties that mark their individual rates of change, they can be partially correlated (cross-correlation), i.e., they can fade, with more or less probability, at the same locations.

Here, we want to simulate such partially correlated signals. A model for the cross-correlation between different received signals was developed by Graziano [10] (and discussed in detail in [9]), where this parameter was found to be dependent on the path angular separation, the distance and the correlation distance, i.e.,

$$\rho = \begin{cases} \sqrt{\dfrac{d_1}{d_2}} & \text{for } 0 \le \phi < \phi_\text{T} \\ \dfrac{\phi}{\phi_\text{T}} \sqrt{\dfrac{d_1}{d_2}} & \text{for } \phi_\text{T} \le \phi < \pi \end{cases} \tag{3.17}$$

where d_1 is the shortest of the two paths: d_1 and d_2. Parameter ϕ_T is given by

$$\phi_\text{T} = 2 \sin^{-1}[L_\text{corr}/(2d_1)] \tag{3.18}$$

where L_corr is the correlation distance.

In this chapter, we are going to generate partially correlated signals using a statistical method based on the Cholesky factorization of the correlation coefficient matrix. This is a statistical technique. In Chapter 2 we have been able to generate partially correlated series using a physical method (Project 2.4).

The measure of the dependence between two or more random variables is given by the covariance. We denote it by C_{ij} or equivalently by $\text{Cov}(X_i, X_j)$ which is defined as

$$C_{ij} = E[(X_i - \mu_i)(X_j - \mu_j)] = E(X_i X_j) - \mu_i \mu_j \tag{3.19}$$

where X_i and X_j with $i = 1, 2, \ldots, n$ and $j = 1, 2, \ldots, n$ are random variables, with $E[]$ being the expectation operator.

Some important properties of the covariance are as follows:

- The covariance is symmetric, i.e., $C_{ij} = C_{ji}$.
- If $i = j$, then $C_{ij} = C_{ji} = \sigma_i^2$, where σ_i is the standard deviation of X_i.
- The covariance can take up positive and negative values.

When $C_{ij} = 0$, the random variables X_i and X_j are said to be *uncorrelated*. If X_i and X_j are independent, their covariance, $C_{ij} = 0$. This does not mean that, if the covariance is zero, then they are independent. However, if X_i and X_j follow a joint Gaussian distribution with $C_{ij} = 0$, then they are also independent.

The difficulty in using the covariance as a measure of the dependence between two random variables comes from the fact that it is not nondimensional. Another parameter, easier to understand is used, the *correlation coefficient*, denoted by ρ_{ij} and defined as

$$\rho_{ij} = \frac{C_{ij}}{\sqrt{\sigma_i^2 \sigma_j^2}} i = 1, 2, \ldots, n; \ j = 1, 2, \ldots, n \tag{3.20}$$

This parameter gives a measure of the linear dependence between variables X_i and X_j. Moreover, its sign is the same as that of the covariance, C_{ij}, and takes up values in the range: $-1 \le \rho_{ij} \le 1$.

We are interested in producing series that, while they follow a Gaussian distribution, they also keep a given correlation between them. The algorithm that we will use follows the steps below:

- Generate identically distributed series of the same length, $Z \sim N(0,1)$.
- Build the covariance matrix, Cov. The correlation coefficient matrix can also be used and later correct the resulting series to achieve the wanted means and standard deviations.
- Perform the decomposition of matrix $\text{Cov} = CC^T$, where C is a lower triangular matrix.
- Finally, compute

$$\boldsymbol{X} = \boldsymbol{\mu} + \boldsymbol{CZ} \tag{3.21}$$

with X containing the resulting signals with the specified cross-correlations.

We start off by using zero mean and unit Gaussian distributions, $N(0,1)$, and later we will include the means and standard deviations: given a normal random variable $X \sim N(0,1)$, it is possible to obtain another random variable, such that $X' \sim N(\mu, \sigma^2)$, by simply performing the operation $X' = \mu + \sigma X$.

The covariance matrix for variables $X_1, X_2, \ldots, X_m$ is given by,

$$\text{Cov} = \begin{pmatrix} \sigma_1^2 & \rho_{12}\sigma_1\sigma_2 & \rho_{13}\sigma_1\sigma_3 & \cdots & \rho_{1m}\sigma_1\sigma_m \\ \rho_{12}\sigma_1\sigma_2 & \sigma_2^2 & \rho_{23}\sigma_2\sigma_3 & \cdots & \rho_{2m}\sigma_2\sigma_m \\ \rho_{13}\sigma_1\sigma_3 & \rho_{23}\sigma_2\sigma_3 & \sigma_3^2 & \cdots & \rho_{3m}\sigma_3\sigma_m \\ \vdots & \vdots & \vdots & \ddots & \vdots \\ \rho_{1m}\sigma_1\sigma & \rho_{2m}\sigma_2\sigma_m & \rho_{3m}\sigma_3\sigma_m & \cdots & \sigma_m^2 \end{pmatrix} \tag{3.22}$$

where the symmetry, $\rho_{ij} = \rho_{ji}$, has been introduced. Assuming unit standard deviations, we would have

$$\text{Cov} = \begin{pmatrix} 1 & \rho_{12} & \rho_{13} & \cdots & \rho_{1m} \\ \rho_{12} & 1 & \rho_{23} & \cdots & \rho_{2m} \\ \rho_{13} & \rho_{23} & 1 & \cdots & \rho_{3m} \\ \vdots & \vdots & \vdots & \ddots & \vdots \\ \rho_{1m} & \rho_{2m} & \rho_{3m} & \cdots & 1 \end{pmatrix}. \tag{3.23}$$

Next the Cholesky factorization of matrix Cov is performed whereby lower triangular matrix C is obtained such that $\text{Cov} = CC^{\text{T}}$. Writing Equation (3.21) in matrix notation, we have

$$\underbrace{\begin{pmatrix} Z_1 \\ Z_2 \\ Z_3 \\ \vdots \\ Z_m \end{pmatrix}}_{\text{Correlated series}} = \underbrace{\begin{pmatrix} 1 & 0 & 0 & \cdots & 0 \\ a_{21} & a_{22} & 0 & \cdots & 0 \\ a_{31} & a_{32} & a_{33} & \ddots & \vdots \\ \vdots & \vdots & \vdots & \ddots & 0 \\ a_{m1} & a_{m2} & a_{m3} & \cdots & a_{mm} \end{pmatrix}}_{C} \underbrace{\begin{pmatrix} X_1 \\ X_2 \\ X_3 \\ \vdots \\ X_m \end{pmatrix}}_{X \sim N(0,1)} \qquad (3.24)$$

For performing the Cholesky factorization MATLAB® function chol can be used. Operating further, the following expressions can be obtained,

$$\begin{aligned} Z_1 &= X_1 \\ Z_2 &= a_{21}X_1 + a_{22}X_2 \\ &\cdots \\ Z_m &= a_{m1}X_1 + a_{m2}X_2 + \cdots + a_{mm}X_m \end{aligned} \qquad (3.25)$$

For the specific case where two random variables are used, we would have

$$\begin{pmatrix} Z_1 \\ Z_2 \end{pmatrix} = \begin{pmatrix} \mu_1 \\ \mu_2 \end{pmatrix} + \begin{pmatrix} \sigma_1 & 0 \\ \sigma_2\,\rho_{12} & \sigma_2\sqrt{1 - \rho_{12}^2} \end{pmatrix} \begin{pmatrix} X_1 \\ X_2 \end{pmatrix} \qquad (3.26)$$

In project32 we use the same basic settings as in previous simulators, but we consider two received signals with equal mean ($-70\,\text{dBm}$) and equal standard deviation (8 dB). The signals have a correlation distance of 30 m while their cross-correlation coefficient matrix is given by

$$\rho = \begin{pmatrix} 1 & 0.9 \\ 0.9 & 1 \end{pmatrix} \qquad (3.27)$$

Figure 3.13(a) illustrates the originally generated, uncorrelated Gaussians. By applying the above method, we can generate partially correlated signals as shown in Figure 3.13(b) where great similarity can be observed between the series, in accordance with the set cross-correlation coefficient value of 0.9.

One final calculation will be performed. We will assume that the two signals correspond to two BSs involved in *soft handover* with the one MS. This is illustrated in Figure 3.14. Soft handover is a very common technique used in spread spectrum systems whereby an MS can simultaneously be connected to two or more BSs for achieving *macrodiversity* and, thus, overcoming shadowing effects. Here, we want to evaluate the coverage probability for the soft-HO scheme when the signals are completely uncorrelated, and when they are partially correlated.

To carry out this calculation, we look for the route points for which, either received signal or both, are above a given operation threshold, set here to $-80\,\text{dBm}$. As in previous simulations, this can be done using MATLAB® function find.

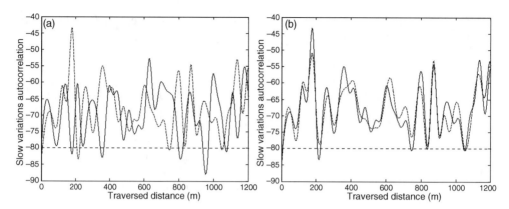

Figure 3.13   (a) Uncorrelated series. (b) Correlated series

Some non-graphical results shown on the MATLAB® screen are as follows for the uncorrelated case,

```
Coverage probability Signal 1. CovProb1=0.9592
Coverage probability Signal 2. CovProb2=0.8722
Coverage probability with diversity Signals 1 & 2. CovProb12=0.9915.
```

For the correlated case, the actually achieved correlation coefficient matrix was

```
1.0000 0.8636
0.8636 1.0000
```

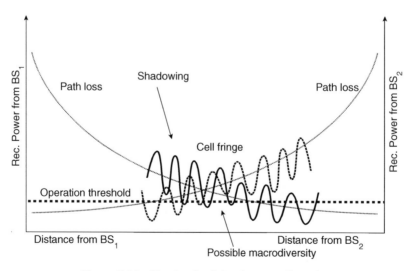

Figure 3.14   Simulated soft handover configuration

which is fairly close to the wanted value, 0.9. In this case, the coverage probabilities were

```
Coverage probability Signal 1. CovProb1=0.9592
Coverage probability Signal 2. CovProb2=0.9360
Coverage probability with diversity Signals 1&2. CovProb12=0.9719.
```

This shows how the diversity improvement is substantially hindered by the existence of a high degree of correlation between the signals involved in the soft HO, i.e., they tend to fade at the same locations.

A simpler calculation could be performed as follows. If the outage probabilities for the two individual BSs are p_1 and p_2, the combined availability would be $a_T = 1 - p_1 p_2$ if both signals were uncorrelated. If some degree of correlation exists, the availability must be calculated in a slightly different way, using the expression,

$$a_T = 1 - \left[\rho \sqrt{p_1(1 - p_1)} \sqrt{p_2(1 - p_2)} + p_1 p_2 \right] \quad (3.28)$$

The same simulation carried out for the two BSs case could be performed for a larger number of BSs. The simulation would become more involved, but would still be easy to implement. This is left for the reader to program.

Project 3.3: Area Coverage

This simulation tries to reproduce a coverage map for one BS. This means simulating both the path loss and the location variability corresponding to the slow signal variations. Such slow variations, which are superposed on the distance-dependent path loss, cannot change abruptly. This means that, in addition to the location variability parameter, σ_L (dB), a correlation distance, L_{corr} (m), must be imposed.

To implement the simulator (project33) the same techniques used in the above projects can be employed. The only difference is that now we have assumed a given correlation length, L_{corr}, that is valid in the x and y coordinates. Thus, we generated independent zero mean and unit standard deviation samples on a square grid of side L_{corr} m. Then, correlated samples were generated in between. To do this, we could perform one-dimensional interpolations row-wise and then column-wise using MATLAB® function interp1. Another alternative is using MATLAB® function interp2 with option spline. This is the option followed in project33.

Afterward, the interpolated samples were multiplied by the location variability and, to the overall interpolated grid, we added the average received power computed using a Hata-like model with a 1-km loss of AA=137 and an exponent nn=3.5; the EIRP was 0 dBm (1 mW). Other settings were: correlation length Lcorr=30 m, sampling spacing ls=5 m and location variability SL=10 dB.

Figure 3.15 illustrates the grid with an L_{corr} m step with uncorrelated Gaussian samples. Figure 3.16 illustrates the interpolated grid with an ls step. Figure 3.17 shows the average power resulting from the use of the propagation model. Figure 3.18 illustrates the result of adding the average power to the interpolated slow variations. Finally, Figure 3.19 shows a plot using MATLAB® function contour where being within the contour means that the

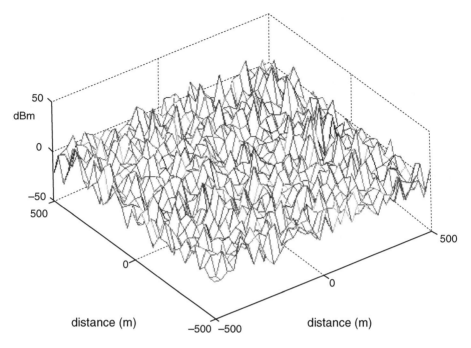

Figure 3.15　Slow signal variability due to shadowing. Coarse, L_{corr}, grid with uncorrelated samples

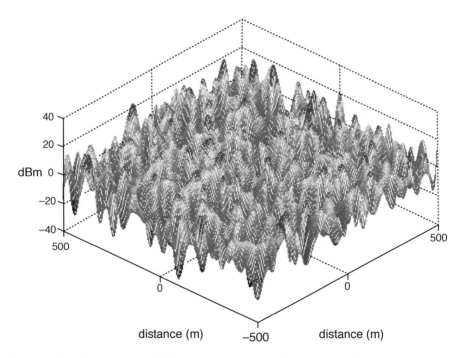

Figure 3.16　Slow signal variability due to shadowing. Fine, d_s, grid with correlated samples

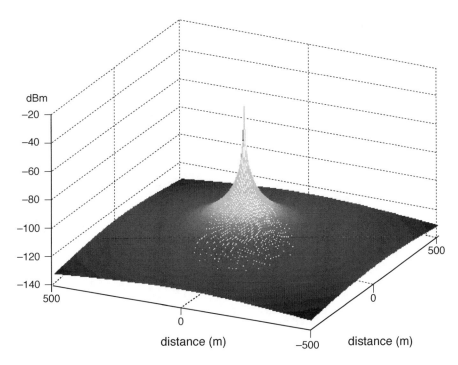

Figure 3.17 Calculated average received power using a Hata-like propagation model. Note the lack of fine detail in the plot

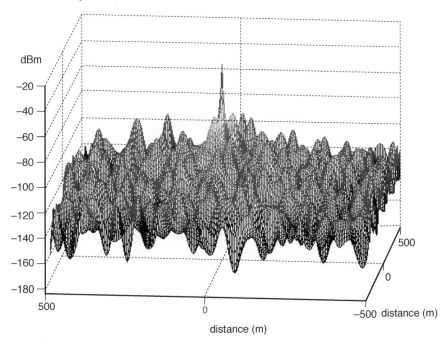

Figure 3.18 Combined plot with average received signal level plus slow variations

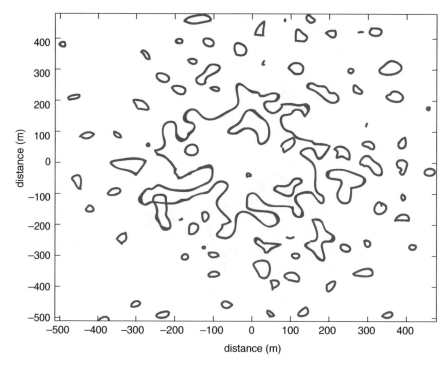

Figure 3.19 Contour coverage plot showing areas above and below threshold where coverage holes are clearly seen

received power is above a given threshold. Note the *coverage holes*. These have to be filled by setting up neighboring BSs and using handover techniques.

In addition to the above plots, we have tried to reproduce the area and fringe coverage calculations mentioned at the beginning of this chapter. We set the fringe at 100 m and we carried out the two calculations by checking whether the received values were above or below a threshold of −110 dBm. The results were as follows:

```
Area coverage probability: 0.98394
Fringe coverage probability:0.95902.
```

Visually compare these figures with the results plotted in Figure 3.19. Note that the range of this BS is too small, this is due to a very low EIRP value and to an exaggeratedly high 1-km loss of 137 dB, thus chosen to artificially make the coverage small in order not to take up a lot of computer memory and computer time.

Project 3.4: Multiple Interference

In this project we try and implement the calculation of the *carrier-to-interference ratio* (CIR) in a multiple interference scenario where the wanted and the interfering signals are

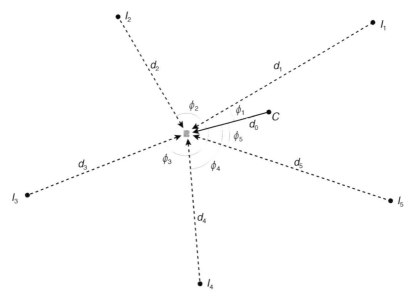

Figure 3.20   Simulation scenario in Project 3.4

partially correlated. The scenario simulated is shown in Figure 3.20 where the wanted signal, C, and five interferers, $I_1, \ldots I_5$, arrive at the receiver. We performed a simple Monte Carlo simulation where we randomly drew values for wanted and interfering signals from their corresponding Gaussian distributions. We accumulated the CIR value for each outcome and, finally, computed its statistics, including the CDF.

To perform the calculation of the overall interference in dBm, we need to convert the interference power from dBm to mW, carry out the sum, and convert the overall interference back to dBm. Thus, the total interference power for one Monte Carlo draw is given by

$$I_{\text{Tot}} = 10 \log(i_1 + i_2 + i_3 + \cdots) \tag{3.29}$$

A total of 5000 draws have been made for calculating the results below. The inputs were those in Table 3.1 where the mean and standard deviation for C and the five interferers are given; in Table 3.2 the assumed correlation matrix is provided, which must be positive definite.

Table 3.1   Input means and standard deviations

Signal	Average power (dBm)	Standard deviation (dB)
C	−80	8
I_1	−110	7
I_2	−120	6
I_3	−125	8
I_4	−108	10
I_5	−90	7

Table 3.2 Matrix of correlation coefficients

ρ_{ij}	C	I_1	I_2	I_3	I_4	I_5
C	1.0	0.5	0.7	0.0	0.0	0.4
I_1	0.5	1.0	0.0	0.0	0.0	0.3
I_2	0.7	0.0	1.0	0.0	0.0	0.0
I_3	0.0	0.0	0.0	1.0	0.5	0.0
I_4	0.0	0.0	0.0	0.5	1.0	0.3
I_5	0.4	0.3	0.0	0.0	0.3	1.0

Figure 3.21(a) shows part of the realizations for C, I_{Tot} and CIR for the uncorrelated case. Also for this case, Figure 3.21(b) shows part of the individual and the overall interference values. Figure 3.22 shows the same plots for the correlated case. Figure 3.23 shows the CDFs of the CIRs for both cases, where it can be clearly seen how the existence of a given amount of correlation impacts the overall CIR statistics. Finally, the CIR's mean and standard deviation for both cases were calculated:

```
CIR mean: 9.0427 dB
CIR std: 10.1957 dB (Uncorrelated case)

CIR mean: 9.3379 dB
CIR std: 8.1426 dB (Correlated case)
```

Although not shown, in both cases, the resulting CDFs could be fitted to a Gaussian distribution (project34).

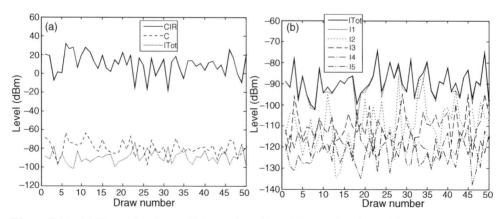

Figure 3.21 (a) Uncorrelated case. C, I_{Tot} and resulting CIR samples. (b) Individual interferers and overall interference. Only 50 out of a total of 5000 draws are shown

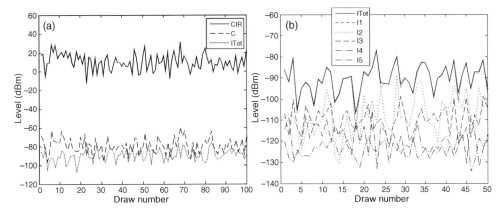

Figure 3.22 Correlated case. (a) C, I_{Tot} and resulting CIR samples. (b) Individual interferers and overall interference. Only 50 out of a total of 5000 draws are shown

3.4 Summary

In this chapter, we have looked into the shadowing phenomenon from a statistical point of view. In Chapter 2 we looked at it from a deterministic point of view. We have tried to provide a number of simulation cases where the various aspects of shadowing are presented, thus, we have first looked at the normal distribution of shadowing when expressed in dB and we have linked the mean of the distribution to the path loss. We have further defined two additional concepts, the location variability and the correlation length. We have presented alternative ways of reproducing such variations. We then went on to present the effect of shadowing cross-correlation, and how to introduce it in simulated series using a statistical approach. In Chapter 2 we were able to generate correlated series using a deterministic

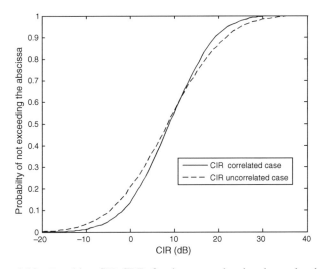

Figure 3.23 Resulting CIR CDFs for the uncorrelated and correlated cases

approach. Finally, we introduced the multiple interference case. We have also used the generated series in hard- and soft-handover examples. Next, we present the multipath phenomenon.

References

[1] B. Sklar. *Rayleigh Fading Channels. Mobile Communications Handbook* (Ed. S.S. Suthersan). CRC Press, 1999.

[2] J.M. Hernando & F. Pérez-Fontán. *An Introduction to Mobile Communications Engineering.* Artech House, 1999.

[3] W.C. Jackes. *Microwave Mobile Communications.* John Wiley & Sons, Inc., New York, 1974.

[4] T.S. Rappaport. *Wireless Communications. Principles and Practice.* Prentice Hall, 1996.

[5] M. Hata. Empirical formula for propagation loss in land mobile radio services. *IEEE Trans. Veh. Tech.*, **29**(3), 1980, 317–325.

[6] Y. Okumura *et al.* Field strength variability in VHF and UHF land mobile service. *Rev. Elec. Comm. Lab.*, Sep–Oct, 1968.

[7] Radiowave Propagation Effects on Next-Generation Fixed-Service Terrestrial Telecommunications Systems. Commission of European Communities. COST 235 Final Report, 1996.

[8] J. Walfisch & H.L. Bertoni. A theoretical model of UHF propagation in urban environments. *IEEE Trans. Anten. Propag.*, **36**(12), 1988, 1788–1796.

[9] S.R. Saunders. *Antennas and Propagation for Wireless Communication Systems.* John Wiley & Sons, Ltd., Chichester, UK, 1999.

[10] V. Graziano. Propagation correlation at 900MHz. *IEEE Trans. Veh. Tech.*, **VT-27**, 1978.

Software Supplied

In this section, we provide a list of functions and scripts, developed in MATLAB®, implementing the various projects and theoretical introductions mentioned in this chapter. They are the following:

```
intro31              GaussianCDF
intro32              fCDF
project311
project312
project313
project314
project32
project33
project34
```

4

Introduction to Multipath

4.1 Introduction

In this chapter, we deal with very simple scenarios for getting acquainted with the multipath phenomenon. An unmodulated RF carrier (continuous wave, CW) is assumed to be transmitted, and basic phenomena, such as the *Doppler* and the *fading*, suffered by the signal will be reproduced through simulations.

It is very important to always bear in mind the geometry of the propagation scenario that is being simulated, since the directions of arrival of the various multipaths have a direct impact on the Doppler spectrum, and on the fade rate. In the simulation projects proposed in this chapter, we will try and reproduce the relationship that exists between the Doppler shift affecting each echo, and their angles of arrival with respect to the mobile's direction of travel.

As in previous chapters, the mobile terminal (MS) is assumed to travel along a straight route. To produce series, the route is tightly sampled (a fraction, F, of the wavelength). Each value in the simulated series will correspond to a route sampling point. Assuming a constant mobile speed, V, the generated series can be represented either in the traveled distance domain (abscissas in meters or wavelengths) or in the time domain (abscissas in seconds). The conversion is simple, i.e., $x = Vt$, where x is the traveled distance.

In the multipath propagation channel, the transmitted signal arrives at the receiver through several echoes due to reflection, diffraction and scattering, mechanisms related, in general, to features such as walls, rooftops, lamp posts, cars, trees, people, etc., in the near environment of the mobile. In Chapter 7 we will deal with *wideband effects*, mainly due to significant multipath contributions farther away from the receiver, i.e., with longer delays.

Multipath is significant in mobile communications, since the MS antenna is usually located at low heights and surrounded by obstacles. It typically uses omnidirectional (omni-azimuth) patterns thus picking up large numbers of echoes. Conversely, typical microwave fixed radio links are sited above the surrounding clutter, and use very directive antennas that discriminate against most of the existing multipath. Only in very specific conditions where anomalous refraction exists, rays will bend down toward the ground giving rise to multipath.

Modeling the Wireless Propagation Channel   F. Pérez Fontán and P. Mariño Espiñeira
© 2008 John Wiley & Sons, Ltd

The unmodulated RF signal can be represented in complex form [1], as discussed in Chapter 1, by

$$a_0 \exp[\mathrm{j}(\omega_c t - k_c z + \phi_0)] \tag{4.1}$$

where a_0 is the magnitude of the direct signal, and ϕ_0 is the initial phase. The term, k_c, is the propagation constant for carrier frequency, f_c, with $k_c = 2\pi/\lambda_c$. The real RF signal is

$$a_0 \cos(\omega_c t - k_c z + \phi_0) \tag{4.2}$$

We can arrive at the low-pass equivalent representation by dropping the carrier term, f_c, in the complex representation, i.e.,

$$a_0 \exp[-\mathrm{j}(k_c z - \phi_0)] \tag{4.3}$$

For mobile channels where the terminal is on the move, the mobile speed, V, must be taken into account. In some instances, the speed of the other terminal must also be introduced, as in the case of *inter-vehicle communications*, or in *land mobile satellite* (LMS), links with non-geostationary orbit (non-GEO) satellites (Chapter 9).

4.2 Projects

After the above introduction, we start our study of the multipath effects by introducing very simple scenarios, and gradually increase their complexity.

Project 4.1: Direct Signal, Angle of Arrival and Doppler Shift

Here, we simulate a link between a fixed transmitter or base station (BS) and a mobile receiver (MS). It is assumed that only the direct signal is present, i.e., no scatterers introduce further echoes [2]. The simulation is performed for a short stretch of the mobile route, thus the received signal has a constant amplitude, but presents a rapidly varying phase. At the same time, the received RF carrier frequency is shifted (Doppler effect).

Next, we analyze the implications of the movement of MS. We simulate the case where BS transmits and MS receives, i.e., we study the *downlink*. Identical phenomena can be observed in the *uplink*.

The simulation we want to carry out is implemented with reference to Figure 4.1, where BS is located somewhere on the x-axis at a distance D, i.e., it is located at point $x = D$. The MS route starts at the origin of coordinates. If MS moves forward, the sampling points are

$$\begin{aligned} x[0] &= 0, \\ x[1] &= \Delta x, \\ x[2] &= 2\Delta x, \\ &\cdots \end{aligned} \tag{4.4}$$

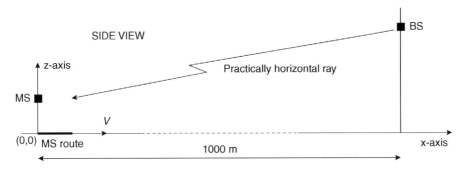

Figure 4.1 Simulation geometry for Project 4.1. Side view

where Δx is the sampling interval, which should be sufficiently small, i.e., a fraction, F, of the wavelength. On the other hand, the associated time samples, $t[n]$, which exactly correspond to the same route sampling points, $x[n]$, are given by

$$
\begin{aligned}
t[0] &= x[0] \ V = 0 \ V, \\
t[1] &= x[1] \ V = \Delta x \ V, \\
t[2] &= x[2] \ V = 2\Delta x \ V,
\end{aligned}
\tag{4.5}
$$

$$\cdots$$

where a constant mobile speed, V, has been assumed. It is also important to program the calculation of the BS–MS distance vector, in this case,

$$
\begin{aligned}
d[0] &= D, \\
d[1] &= D - \Delta x, \\
d[2] &= D - 2\Delta x,
\end{aligned}
\tag{4.6}
$$

$$\cdots$$

Note how the BS–MS antenna height difference was supposed to be sufficiently small so that we can assume a horizontal link (Figure 4.1), thus reducing the problem to a two-dimensional (2D) one.

We also assume that, for a small simulation route, the amplitude of the direct signal does not change while the phase does. Thus, the complex, low-pass equivalent received signal envelope is

$$
\begin{aligned}
r[0] &= \exp(-\mathrm{j}k_c d[0]), \\
r[1] &= \exp(-\mathrm{j}k_c d[1]), \\
r[2] &= \exp(-\mathrm{j}k_c d[2]),
\end{aligned}
\tag{4.7}
$$

$$\cdots$$

where a normalized amplitude equal to one has been assumed. As for the initial phase, it has been set to zero. With this, the simulation is completed. This is the extremely simple outcome of our first simulator in this chapter. Now, what remains is calculating and plotting its amplitude, phase and spectrum.

In this simulation (`project411`) we have assumed a BS–MS distance, $D = 1$ km, with MS approaching BS (Figure 4.1). The mobile route was of 100 samples, the working frequency was 2 GHz (3G systems). The route was sampled with a spacing $\Delta x = \lambda/16$ (F=16). The mobile speed was $V = 10$ m/s (36 km/h).

The corresponding wavelength is $\lambda_c = c/f_c = 0.15$ m, where c is the speed of light, 3×10^8 m/s. The sampling interval is $\Delta x = \lambda/16 = 0.009375$ m $= 9.375$ mm. This is in the traveled distance domain; in the time domain, the sampling interval, t_s, depends on the mobile speed, V, thus $t_s = \Delta x/V = 0.0009375$ s which is equivalent to a sampling frequency, $f_s = 1/t_s = V/\Delta x = 1067$ Hz.

With the above settings, this simulation corresponds to a route length of 100×0.15 m$/16 = 0.9375$ m which, traversed at a speed of 10 m/s, corresponds to a simulation time of 0.9375 m$/10$ m/s $= 0.09375$ s. When plotting the simulated series, $r[n]$, this can be referred either to time, $t[n]$, or to traveled distance, $x[n]$.

Now, we go on to plot the various parameters of the complex envelope, $r[n]$. Its magnitude, $|r|$, is obviously constant and equal to one. The phase, however, changes with time and/or locations. To calculate the phase MATLAB$^{\circledR}$ (MATLAB$^{\circledR}$ is a registered trademark of The MathWorks, Inc.) function `angle` is used. Function `angle` provides the modulo-π phase. To calculate the absolute phase, function `unwrap` can be applied to `angle`. Figure 4.2 shows both the absolute phase as a function of the traveled distance, and the modulo-π phase as a function of the traveled distance.

To calculate the spectrum of the received complex envelope, $r[n]$, MATLAB$^{\circledR}$ function `fft` can be used. The spectrum of the complex envelope can be computed by taking the square of the magnitude of its Fourier transform. In MATLAB$^{\circledR}$ code, this can be implemented using

```
(abs(fft(r,N)).^2)/N
```

or, alternatively,

```
(fft(r,N).*^conj(fft(r,N)))/N
```

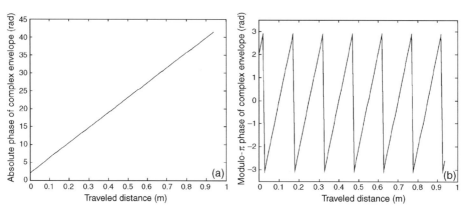

Figure 4.2 (a) Simulated absolute phase and (b) modulo-π phase as a function of the traveled distance

where N is the number of points in the FFT. In our implementation, we normalize the computed spectrum with respect to its maximum.

It is important to properly set the frequency axis. MATLAB® calculates the FFT so that the first element of the resulting vector corresponds to frequency 0 Hz, the other elements going all the way to the sampling frequency, f_s. Using MATLAB® function `fftshift`, it is possible to shift around the resulting vector so as to place, at the center of the vector, the result corresponding to 0 Hz. Thus, we have to be very careful in setting the frequency axis right. In MATLAB® code, this could be done as follows:

```
freqaxis=(-(N/2):(N/2)-1)*fs/N
```

where N is the number of FFT points.

In Figure 4.2 it can clearly be observed how the phase, both absolute and modulo-π, increases as MS travels toward BS. Note that the distance decreases, but the phase increases given the negative sign in front of the phase term. As for the spectrum, Figure 4.3 shows a clear line but no longer at 0 Hz (which represents the transmitted carrier frequency) but at a positive frequency, i.e., the transmitted CW carrier has been shifted (Doppler effect).

It is important to bear in mind that the amplitude of the direct signal does not vary significantly provided that the route is short enough. Coming back to our simulation, MS is 1 km away from BS and the simulation route is of 100 samples spaced λ/F with MS approaching BS (Figure 4.1). Given that the frequency is 2000 MHz, which corresponds to a wavelength of 0.15 m, the radio path will be 1000 m long for the first route sampling point, and $1000 - 100 \times 0.15/16 = 999.0625$ m for the last. The power ratio at these two distances, assuming free-space propagation conditions, is

$$\frac{p_{0\,@1000\,\mathrm{m}}}{p_{0\,@999.0625\,\mathrm{m}}} = \frac{a_{0\,@1000\,\mathrm{m}}^2}{a_{0\,@999.0625\,\mathrm{m}}^2} = \left(\frac{1000}{999.0625}\right)^2 \approx 1.0019 \equiv 0.0081\,\mathrm{dB} \qquad (4.8)$$

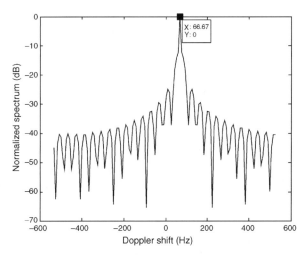

Figure 4.3 Doppler spectrum of low-pass equivalent signal

which is a negligible difference. Hence, the same magnitude can be used for the direct signal throughout the simulation route (tens to hundreds of wavelengths). If the BS–MS distance is longer, the above difference will become even smaller. This will not be the case with the phase, i.e., small radio path differences will mean several 2π 'turns' in the phase term, $\exp(-jk_c d[n])$, where $d[n]$ indicates distance sample number.

Now, we try to explain with simple mathematics the simulated results. If MS approaches BS at a speed V, the received complex envelope, $r(t)$, is

$$r(t) = a_0 \exp[-jk_c d(t)] = a_0 \exp\left[-j\frac{2\pi}{\lambda_c}d(t)\right] \qquad (4.9)$$

where, for convenience, the initial phase, ϕ_0, has been dropped. The term $d(t)$ is the BS–MS distance (radio path) that varies according to V. As said above, as $d(t)$ increases the phase decreases and vice versa, given the negative sign in front of the phase term. To show this explicitly, let us put the distance as a function of its initial value plus a time-varying term, i.e.,

$$d(t) = d_0 - Vt \qquad (4.10)$$

The received complex envelope then becomes

$$r(t) = a_0 \exp\left(-j\frac{2\pi}{\lambda_c}d_0 + j\frac{2\pi}{\lambda_c}Vt\right) = a_0 \exp\left(j\xi + j\frac{2\pi}{\lambda_c}Vt\right) \qquad (4.11)$$

Again, dropping for convenience the constant phase term, we get

$$r(t) = a_0 \exp\left(j\frac{2\pi}{\lambda_c}Vt\right) = a_0 \exp\left[j2\pi\left(\frac{V}{\lambda_c}\right)t\right] = a_0 \exp(j2\pi f_D t) \qquad (4.12)$$

where f_D is the *Doppler shift* with respect to the nominal carrier frequency represented, in the low-pass equivalent domain, by 0 Hz. We can come back to the phasor representation of the received RF signal by including the carrier, $a_0 \exp[j2\pi(f_c + f_D)t]$, and to its real RF representation, by taking the real part, i.e., $a_0 \cos[2\pi(f_c + f_D)t]$.

It can clearly be observed how the CW carrier frequency is no longer received at its nominal value but at $f_c + f_D$. The shift is $f_D = V/\lambda_c$ Hz. If MS traveled away from BS, the phase would decrease, thus giving rise to a negative Doppler.

Next, two more simulations are proposed. The first one corresponds to MS traveling away from BS (project412) ($\alpha = 180°$). The other simulation assumes an angle, $\alpha = 120°$, of the radio link with respect to the mobile route (project413). In this case, BS is located at coordinates D_x and D_y, at a distance $D = 1000$ m from the origin, that is, from the first point of the MS route. Figure 4.4 shows how the phase decreases in the $\alpha = 180°$ case, it also decreases in the $\alpha = 120°$ case, but with a gentler slope. Similarly, Figure 4.5 presents their corresponding spectral lines which reflect the different slopes.

It can be shown that the following relationship, between the angle of arrival, α, and the Doppler shift, is fulfilled,

$$f_D(\alpha) = f_{max}\cos(\alpha) = \frac{V}{\lambda_c}\cos(\alpha) \qquad (4.13)$$

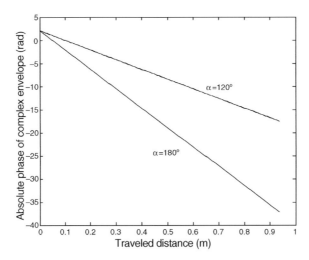

Figure 4.4   Absolute phase when MS travels away from BS with $\alpha = 180°$ and $\alpha = 120°$

Of course, when the radio path is perpendicular to the MS route, i.e., $\alpha = 90°$, $f_D(\alpha = 90) = 0\,Hz$.

We now try to explain Equation 4.13. We assume that rays from BS or from a point scatterer remain parallel for all points of the mobile route (Figures 4.6 and 4.7). This is approximately true for short route lengths. The radio path increment for each route sampling point is given by

$$\Delta l = -\frac{\lambda}{F}\cos(\alpha) \tag{4.14}$$

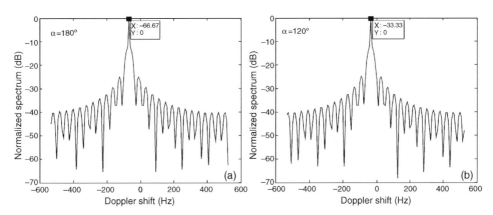

Figure 4.5   (a) Doppler spectrum when MS travels away from BS ($\alpha = 180°$), the shift is $-V/\lambda_c$ $\cos(180) = -66.67\,Hz$. (b) And $\alpha = 120°$, the shift is $-V/\lambda_c \cos(120) = -33.33\,Hz$

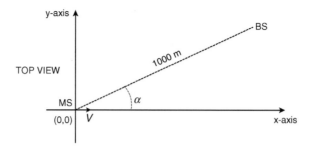

Figure 4.6 Project 4.1 geometry. Top view

where, as indicated earlier, F (fraction), is the number of samples per wavelength. Consequently, the phase increment, is given by

$$\Delta\phi = -\frac{2\pi}{\lambda}\Delta l = \frac{2\pi}{\lambda}\frac{\lambda}{F}\cos(\alpha) = \frac{2\pi}{F}\cos(\alpha) \tag{4.15}$$

For $|\alpha| < 90°$, we will have positive phase increments (positive Dopplers), while for $\alpha = \pm90°$ the phase increment will be zero and so will the Doppler. Finally, for $|\alpha| > 90°$ the phase decreases and the Doppler will be negative.

Bearing in mind that the relationship between phase and frequency is given by

$$f_D = \frac{1}{2\pi}\frac{d\phi(t)}{dt} \approx \frac{1}{2\pi}\frac{\Delta\phi(t)}{\Delta t} \tag{4.16}$$

as $\Delta x/\Delta t = \Delta x/t_s = V$, then

$$f_D \approx \frac{1}{2\pi}\frac{\Delta\phi(t)}{\Delta t} = \frac{1}{2\pi}\frac{\frac{2\pi}{F}\cos(\alpha)}{\frac{\lambda}{FV}} = \frac{V}{\lambda}\cos(\alpha) \tag{4.17}$$

This expression gives the Doppler shift for a given echo as a function of its angle of arrival.

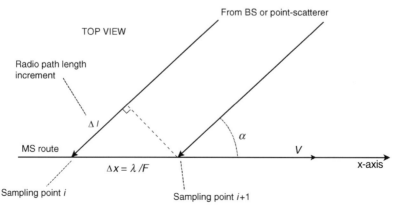

Figure 4.7 Parallel ray assumption

Project 4.2: Direct and Reflected Signals

In this project we will assume that two signals arrive at the receiver: the direct and a reflected ray. In this case, fades already start to appear (`project421`). When two rays interfere with each other in space, they give rise to a *standing wave*, i.e., a pattern of signal enhancements and partial or total cancellations which remain stationary in space. This means that if a probe, i.e., an antenna, senses the resulting signal several times, passing through the same spots and at the same height, the same signal levels will be observed. This standing wave model is realistic but does not totally match reality. Rather, this spatial pattern is, in fact, moving slowly around. Thus a *quasi-stationary pattern* will actually be a more accurate picture of the multipath phenomenon.

For `project421`, we assume that a direct and a reflected signal, e.g., off a building face, exist (Figure 4.8). First, we suppose a reflection coefficient of magnitude one and phase 180°, i.e., $R = -1$.

As illustrated in Figure 4.8, we assume that MS travels toward the reflector and away from BS, all being on the x-axis. The simulation inputs are the same as in Project 4.1. The distances to the origin of both BS and the reflector are 1000 m as shown in the figure. We already should have guessed that two Doppler lines will be present in the received signal spectrum: one positive corresponding to the reflected signal and another negative corresponding to the direct signal.

To perform the simulation, again, sampling points along the mobile route must be defined together with a distance matrix (BS–MS and BS–Reflector–MS), i.e.,

$$
\begin{aligned}
&x[0] = 0, &&d_0[0] = D, &&d_1[0] = D + 2D_{\mathrm{R}}, \\
&x[1] = \Delta x, &&d_0[1] = D + \Delta x, &&d_1[1] = D + 2D_{\mathrm{R}} - \Delta x, \\
&x[2] = 2\Delta x, &&d_0[2] = D + 2\Delta x, &&d_1[2] = D + 2D_{\mathrm{R}} - 2\Delta x, \\
&\cdots &&\cdots &&\cdots
\end{aligned}
\tag{4.18}
$$

Distance d_0 (radio path of the direct ray) increases as MS drives away from BS while d_1 (radio path of the reflected ray) decreases as MS drives toward the reflector. The resulting

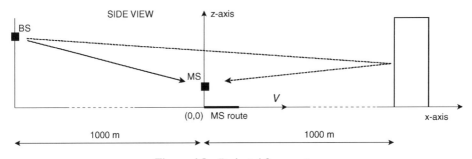

Figure 4.8 Project 4.2 geometry

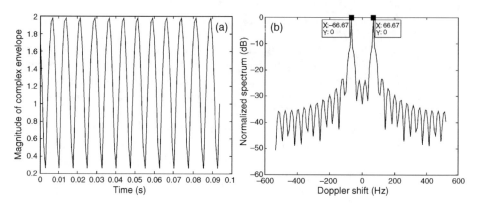

Figure 4.9   (a) Received signal amplitude. (b) Doppler spectrum of received signal

complex envelope is the sum of two phasors, including the reflection coefficient (magnitude and phase), i.e.,

$$
\begin{aligned}
r[0] &= \exp(-\mathrm{j}k_c d_0[0]) - \exp(-\mathrm{j}k_c d_1[0]), \\
r[1] &= \exp(-\mathrm{j}k_c d_0[1]) - \exp(-\mathrm{j}k_c d_1[1]), \\
r[2] &= \exp(-\mathrm{j}k_c d_0[2]) - \exp(-\mathrm{j}k_c d_1[2]),
\end{aligned}
\tag{4.19}
$$

$$\cdots$$

The overall complex series, $r[n]$, is already the outcome of our simulation. Again, what remains is plotting its parameters. In this case, we are interested in its amplitude which is given by MATLAB$^{®}$ function `abs`. Figure 4.9(a) illustrates the resulting amplitude. We can see how the received amplitude fluctuates between 2 ($+6$ dB) and 0 ($-\infty$ dB), i.e., from double the direct signal's amplitude down to total cancellation. Note that, even though we are using a very tight sampling spacing, i.e., $F = 16$, this is not enough to capture in the figure the very deep fades. We can, however, observe that, already with two contributions, fade effects arise. Figure 4.9(b) shows the spectrum of the overall complex envelope. In this spectrum, it is clearly shown how two lines of opposite signs are received, as expected. Their frequencies are $f_{\mathrm{Dir}} = -V/\lambda$ Hz for the direct contribution and $f_{\mathrm{Refl}} = +V/\lambda$ Hz for the reflected contribution.

Again, we can try to explain the obtained results by analyzing the phasors corresponding to the two contributions reaching MS. Thus,

$$
r(t) = \exp\left[-\mathrm{j}\frac{2\pi}{\lambda_c}d_0(t) + \mathrm{j}\phi_0\right] - \exp\left[-\mathrm{j}\frac{2\pi}{\lambda_c}d_1(t) + \mathrm{j}\phi_1\right]
\tag{4.20}
$$

The two radio distances change in time, i.e.,

$$
d_0(t) = D + Vt \text{ and } d_1(t) = D + 2D_{\mathrm{R}} - Vt = 3D - Vt
\tag{4.21}
$$

We can see how d_0 increases while d_1 decreases. Rearranging the complex envelope expression and dropping the initial phases for simplicity,

$$
\begin{aligned}
r(t) &= \exp\left(-j2\pi\frac{V}{\lambda_c}t - j\frac{2\pi}{\lambda_c}D\right) - \exp\left(j2\pi\frac{V}{\lambda_c}t - j\frac{2\pi}{\lambda_c}3D\right) \\
&= \exp\left(-j2\pi\frac{V}{\lambda_c}t + j\phi_0'\right) - \exp\left(j2\pi\frac{V}{\lambda_c}t + j\phi_1'\right) \\
&= \exp\{j[2\pi(-f_D)t + \phi_0']\} - \exp[j(2\pi f_D t + \phi_1')]
\end{aligned}
\tag{4.22}
$$

and dropping again the fixed phase terms, we get

$$
r(t) = \exp\left(-j2\pi f_D t\right) - \exp\left(j2\pi f_D t\right)
\tag{4.23}
$$

Taking into account the equality

$$
\exp(j2A) - \exp(j2B) = 2j\sin(A - B)\exp[j(A + B)]
\tag{4.24}
$$

and equating $2A = -2\pi f_D t$ and $2B = 2\pi f_D t$, we obtain

$$
\begin{aligned}
r(t) &= 2j\sin\left(\frac{-2\pi f_D t}{2} - \frac{2\pi f_D t}{2}\right)\exp\left[j\left(\frac{-2\pi f_D t}{2} + \frac{2\pi f_D t}{2}\right)\right] \\
r(t) &= -2j\sin\left(2\pi f_D t\right)
\end{aligned}
\tag{4.25}
$$

Concentrating now on the magnitude of the complex envelope, i.e.,

$$
|r(t)| = |2\sin(2\pi f_D t)|
\tag{4.26}
$$

we can observe a sinusoidal oscillation (semi-sinusoidal, rather) or *fade pattern* of frequency $2f_D$, because the sine function folds over at the negative values, and takes on values going from double the direct signal amplitude down to total cancellation.

The expression for the RF received signal can be obtained by inserting back the RF carrier term and taking the real part of the RF phasor. Figure 4.10 illustrates this. Note the marked

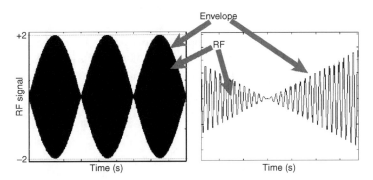

Figure 4.10 Carrier frequency modulated as MS moves through the standing wave pattern created by the interference between the direct and the reflected rays

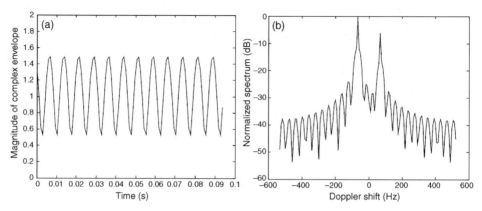

Figure 4.11   (a) Received signal amplitude. (b) Doppler spectrum

difference, in the rates of change, of the modulation introduced by the channel, which has a maximum frequency related to the maximum Doppler shift, and that of the RF frequency.

It is also interesting to study how these periodic, semi-sinusoidal fades behave in the traveled distance domain. By putting the time variable, t, as a function of traveled distance and the mobile speed, we get

$$|r(x)| = \left| 2 \sin\left(\frac{2\pi}{\lambda_c} x\right) \right| \tag{4.27}$$

Thus, the magnitude of the complex envelope will go through two nulls and two maxima within each wavelength of traveled route. Faster rates of change will not be possible. Later, we will look into the case of two moving terminals (Project 4.7) where we continue studying this point.

The simulation is repeated (`project422`) assuming that the reflection coefficient is now $R = -0.5$. In this case, the results shown in Figure 4.11 are obtained. As expected, the maximum value of the magnitude of the complex envelope, i.e., when both echoes are received in phase, is $1 + 0.5 = 1.5$, i.e., $20\log(1.5) = 3.52\,\text{dB}$. The minimum value, corresponding to echoes received in phase opposition, is $1 - 0.5 = 0.5$, i.e., $20\log(0.5) = -6\,\text{dB}$. Again, it can also be observed how the number of cycles per wavelength (0.15 m) is 2, i.e., two maxima and two minima.

In the frequency domain, the spectrum shows two Doppler lines, one with a negative Doppler equal to $-V/\lambda = -66.67\,\text{Hz}$ and equal to a normalized level of 0 dB, while the other line shows a positive Doppler equal to 66.67 Hz, and a relative level of -6 dB, i.e., half the direct ray's amplitude: 20 log (0.5).

Project 4.3: Two Scatterers

Here, two point scatterers are assumed to be the only sources through which the transmitted signal arrives at the receiver, while the direct signal is assumed to be totally blocked. We are interested in relating two aspects of the multipath phenomenon: *fade rate* and *Doppler spread* [2].

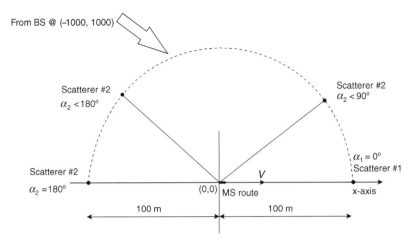

Figure 4.12   Project 4.3 simulation geometry

The specular reflection is not the only mechanism giving rise to multipath, on the contrary, scattering (Chapter 1) is its most common source. Scattering, or diffuse scattering, is generated on small objects, e.g., lamp posts, traffic signals, cars, trees or rough surfaces, e.g., irregular building face surfaces or irregular terrain, giving rise to weaker echoes than those produced on large, flat, smooth surfaces such as the ground or smooth glass building faces that have a specular character. In many simulations throughout the book we will assume that scattered echoes originate at *point scatters*.

The simulation performed in Project 4.2 is repeated here assuming that the direct signal is blocked and two scattered signals arrive at the receiver. First (`project431`), we assume that the point sources of these scattered signals are located in front and behind MS, thus the echoes form two angles $\alpha_1 = 0°$ and $\alpha_2 = 180°$ with MS's motion vector (Figure 4.12). We have to take into consideration the paths between BS and SC#1 and SC#2. However, the Doppler shifts will be dependent on the angles of arrival, α_1 and α_2, since it is only the second sub-paths (SC–MS) that change.

Again, we proceed as before by defining the sampling points along the MS route, $x[n]$, calculating the distance vectors (distance matrix) for each point scatterer and each point along the MS route, $d_1[n]$ and $d_2[n]$, and computing the complex envelope as

$$r[n] = \exp(-jk_c d_1[n]) + \exp(-jk_c d_2[n]) \tag{4.28}$$

where amplitudes equal to one have been assumed for both rays.

Next, we change point scatterer 2 position to form several angles (`project432`, `project433`, `project434`), $30° < \alpha_2 < 180°$, with the traveled route so that the *angle spread*, ξ, reduces with respect to the first case where $\xi = 180°$ down to $\xi = 30°$. Figure 4.13 shows how the received signal pattern presents decreasing *fade rates* (*rates of change* are slower). On the other hand, Figure 4.14 shows the corresponding spectra. It can be observed how the fade rate decreases as the *angle spread*, ξ, decreases. This is accompanied by a reduction in the observed *Doppler spread*, defined here as the difference between the maximum and minimum Doppler components in the received signal. Thus, we see how the

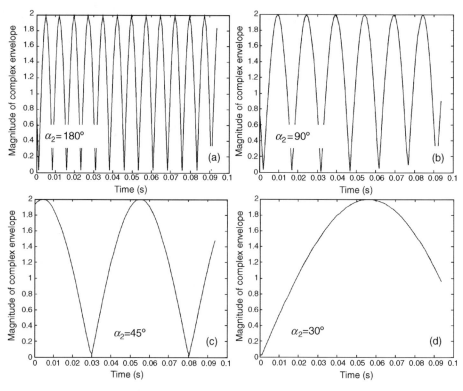

Figure 4.13 Fade patterns for (a) $\alpha_2 = 180°$. (b) $\alpha_2 = 90°$. (c) $\alpha_2 = 45°$. (d) $\alpha_2 = 30°$

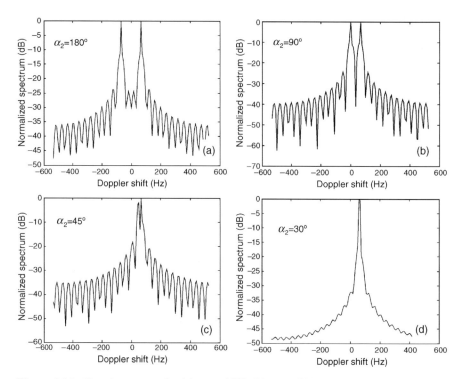

Figure 4.14 Doppler spectra for (a) $\alpha_2 = 180°$. (b) $\alpha_2 = 90°$. (c) $\alpha_2 = 45°$. (d) $\alpha_2 = 30°$

Doppler spread parameter controls the bandwidth of the fading process. As the bandwidth decreases, the fade rate becomes slower. In the limit, when the angle separation is $\xi = 0°$, we return to the single echo case (Project 4.1) where there was no fading.

By developing the sum of the two phasors [2] as in the preceding project, the general expression for the fade rate for two unit-amplitude rays with angles of arrival α_1 and α_2 can be written according to the following formula,

$$|s_r| = \left|2\sin\left\{2\pi\frac{V}{2\lambda}[\cos(\alpha_1) - \cos(\alpha_2)]t\right\}\right| = \left|2\sin\left\{2\pi\left(\frac{f_D}{2}\right)[\cos(\alpha_1) - \cos(\alpha_2)]t\right\}\right| \quad (4.29)$$

that clearly shows how, as the angle spread, i.e., angular sector between α_1 and α_2 decreases, the fade rate decreases as well.

Project 4.4: Multiple Scatterers

Here we go on and extend the use of the point-scatterer model presented in the previous project by including multiple point scatterers (project44). This model will be used extensively throughout the book. In the above simulations simple, periodic series were synthesized. If the number of point scatterers is increased, more intricate time series can be produced but still, they will keep a pseudo-periodic character.

With this multiple point-scatterer assumption (Figure 4.15), it will be possible to simulate many typical situations in mobile propagation such as Rayleigh fading conditions (Chapter 1). The number of rays required to simulate this type of variations is small, with six or seven this should already be possible. To reproduce effects other than the amplitude, a larger number of rays, producing a continuum-like scenario, would be necessary. This is left for the next chapter. The examples reproduced here have been generated using 13 rays. It must be borne in mind that, in most cases, multipath conditions correspond to high densities of rays, i.e., a continuum of rays. In some cases, specific rays may dominate over others, e.g., specular rays dominating over diffuse components with lower magnitudes.

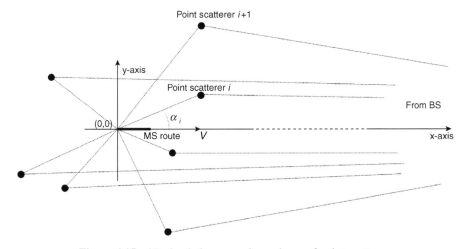

Figure 4.15 2D simulation scenario made up of point scatterers

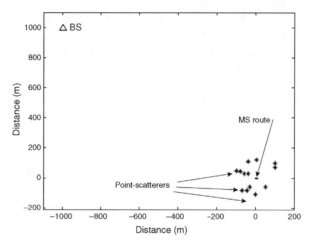

Figure 4.16   Point-scatterer scenario used in the simulated plots

Simulation scenarios, made up of point scatterers can be created by defining their positions by means of random generators (as in Chapter 5). Here, their positions have been set by means of a text editor, and included in the body of the MATLAB$^{®}$ script (`project44`) using a matrix of x and y coordinates. Similarly, the position of BS can be set by specifying its distance to the origin and its angle with respect to the mobile terminal route along the x-axis. As in the previous projects, in order to make the simulation easier, a horizontal 2D geometry is considered, i.e., BS, MS and the various point scatterers are at the same height so that rays are parallel to the ground. Ray magnitudes are assumed to be all equal to one. The same inputs as in previous projects are assumed. The actual simulation geometry is illustrated in Figure 4.16.

As in previous projects, the route sampling points, $x[n]$, and the distance matrix containing all BS–SC–MS distances, $d_i[n]$, are calculated first. Then, the overall complex envelope, $r[n]$, is worked out by computing the complex sum of all point-scatterer phasors. The direct signal is assumed to be totally blocked, i.e., $a_0 = 0$ (NLOS case).

Figure 4.17(a) shows the resulting magnitude of the complex envelope in logarithmic units, $20 \log(|r|)$, and Figure 4.17(b) shows the Doppler spectrum. Figure 4.18 illustrates the

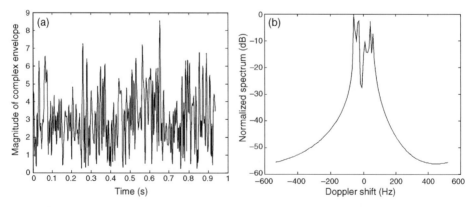

Figure 4.17   (a) Simulated complex envelope magnitude. (b) Doppler spectrum

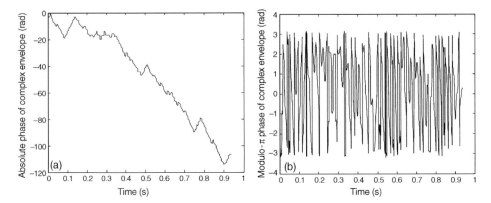

Figure 4.18   (a) Absolute phase. (b) Modulo-π phase of simulated complex envelope

phase. The number of spectral lines in the Doppler spectrum should be equal to the number of multipath components, i.e., 13, unless two or more lines are closer together than the frequency resolution of the FFT, which depends on the number of points used: 64, 128, 256, etc. There is also a left–right ambiguity as $V/\lambda \cos(\alpha) = V/\lambda \cos(-\alpha)$. If a continuum of scatterers were considered, the assumption being closer to reality, individual spectral lines would be indistinguishable.

We are also interested in checking whether the obtained magnitude variations follow a Rayleigh distribution. To verify this, the sample cumulative distribution, CDF, was calculated first (fCDF) as in Chapter 1, and then compared with the theoretical CDF (RayleighCDF). It is possible to link the simulation parameters, namely, the number of rays, N, and the ray amplitudes, a_i, with the Rayleigh distribution parameter, σ. The following expression is fulfilled,

$$\sum_{i=1}^{N} a_i^2 = Na^2 = 2\sigma^2 \tag{4.30}$$

with $2\sigma^2$ being the rms squared value of the Rayleigh distribution. In our case, $N = 13$ and $a_i = 1$, thus, $\sigma = 2.55$. Figure 4.19 shows the simulated and theoretical CDFs. A very good agreement can be observed. Simple goodness-of-fit procedures have already been discussed in Chapter 1.

Project 4.5: Standing Wave due to Multipath

As indicated earlier in this chapter, we are assuming that the multipath conditions present in the area MS is traversing do not change with time. This is an often made assumption that is fairly close to reality although a much better approximation would be that of a quasi-standing wave, slowly changing in time. We have modified the simulator from the previous project to sample a $3 \times 3 \, \text{m}^2$ area to explicitly show the standing wave assumption made.

In scrip project45, to make the representation simpler, we have lowered the carrier frequency to 200 MHz, and increased the sampling rate by using $F = 50$ (fraction of wavelength). Figure 4.20 shows the synthesized spatial distribution of the received signal.

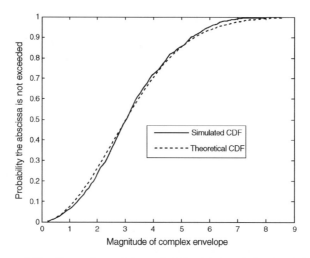

Figure 4.19   Comparison of theoretical Rayleigh and simulated CDFs

Project 4.6: Quasi-standing Wave

Signal variations in multipath environments are not only observed when MS moves, i.e., when it probes a standing wave. If MS is stationary, time variations will be observed. This is due to the fact that some scatterers may be moving, e.g., trees swayed by the wind, pedestrians moving close by, passing cars, etc. Thus, here we alter our stationary multi-point scatterer model for simulating the time variations seen by a static receiver. In the rest of the book, we will stick to the assumption of a time-invariant standing wave.

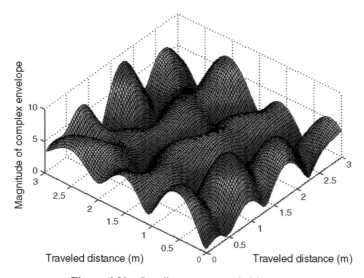

Figure 4.20   Standing wave sampled in space

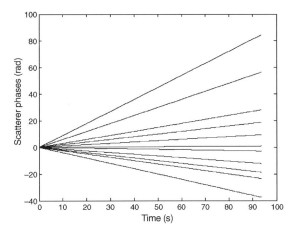

Figure 4.21   Simulated linear phase variations for each scatterer

For simulating their time variations (`project46`), we assume that it is only the phases of the scatterers that change, not their amplitudes. We will model them as simple linear, slowly varying phases. Figure 4.21 illustrates this for the 13 scatterers used in Project 4.4. Now, in addition to the phases introduced by the propagation path distances, additional time-varying phase terms are included, i.e.,

$$r[n] = \sum_{i=1}^{13} \exp(-\mathrm{j}k_c d_i[n] + \mathrm{j}\phi_i[n]) \tag{4.31}$$

The new phase terms are given by $\phi_i[n] = \mathrm{Slope}_i\, t_s[n]$, where Slope_i is a linear phase slope in radians/s. Figure 4.22 shows the magnitude and phase of the simulated complex envelope. When compared with the signal variations obtained in Project 4.4, it can clearly be observed how these variations are much slower. If the Doppler spectrum of the synthesized complex

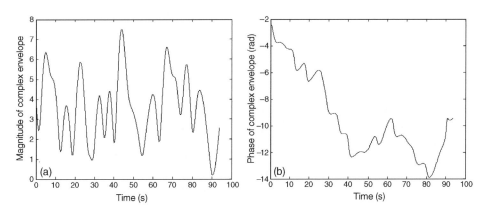

Figure 4.22   (a) Magnitude of the received complex envelope and (b) phase

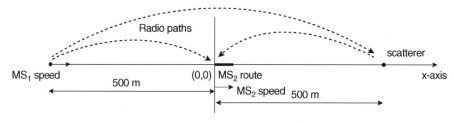

Figure 4.23   Simulated geometry

envelope were plotted, its marked narrowness would be apparent. This exercise is left for the reader to carry out.

Project 4.7: Link Between Moving Terminals

It is not uncommon, in many applications, that radio links are set up between two terminals both on the move. A simple example could be a walkie-talkie communication between two close-by vehicles. More complex links could correspond to modern *intelligent transportation systems* (ITS) where traffic-related data is relayed between vehicles and on to fixed access points in motorways with such communication infrastructures. We will also deal with this type of situation when we review, in Chapter 9, the land mobile satellite channel where links between a mobile terminal and a non-geostationary satellite are set up.

Here, we concentrate on simulating a simple link between two moving MSs. We assume the same frequency as in previous projects. The route length is set to 100 samples spaced $\lambda/16$. The carrier was set to 2 GHz. Let us assume that the initial direct ray path length is 500 m and that only one point scatterer is present (also 500 m away along the x-axis). Figure 4.23 illustrates the simulated geometry.

Again, the simulation process is quite simple, first the sampling points along the route, $x[n]$, are defined. Then the distance matrix is calculated, containing $d_0[n]$ and $d_1[n]$, for the direct and the scattered signals. Finally, the complex envelope is calculated as the sum of two phasors,

$$r[n] = \sum_{i=0}^{1} \exp(-jk_c d_i[n]) \tag{4.32}$$

Amplitudes equal to one have been assumed. The difference with respect to previous projects is that both the transmitter, MS_1, and the receiver, MS_2, can be moving. Even the scatterer could be another car, and be moving as well.

Let us simulate the simple case where the velocities of the two MSs are identical, i.e., $V_1 = V_2 = 10$ m/s, and the scatterer is stationary (`project471`). While distance vector $d_1[n]$ decreases in time, it is clear that $d_0[n]$ does not change, thus the Doppler shift on the direct ray will be zero. On the other hand, the Doppler shift on the scattered signal corresponds to a speed equal to $V_1 + V_2$ as shown in Figure 4.24(a). The maximum Doppler

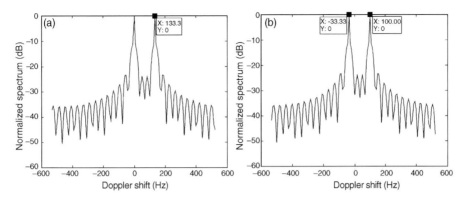

Figure 4.24   Doppler spectra when (a) $V_1 = V_2 = 10\,\text{m/s}$, and (b) $V_1 = 5\,\text{m/s}$ and $V_2 = 10\,\text{m/s}$

shift if MS_1 were stationary would be only $V_2/\lambda = 66.67\,\text{Hz}$ while, in this case, it is $(V_1 + V_2)/\lambda = 133.33\,\text{Hz}$.

If $V_1 = 5\,\text{m/s}$ and $V_2 = 10\,\text{m/s}$ (`project472`), then the Doppler spectrum shown in Figure 4.24(b) is obtained, where the frequency lines correspond to the following Doppler shifts: $(V_1 - V_2)/\lambda = -33.33\,\text{Hz}$ and $(V_1 + V_2)/\lambda = 100\,\text{Hz}$.

The reader is encouraged to try other geometries and velocity combinations, including a moving scatterer. Other suggested simulations could include multiple static and/or moving scatterers.

4.3 Summary

After studying, in previous chapters, the shadowing effects, here we have started the analysis of the multipath phenomenon. This has been introduced in a step-by-step basis by presenting first, very simple geometries, and continuing with more intricate ones. We have seen that multipath is a time-selective phenomenon, i.e., it gives rise to fades. In addition, the movement of at least one of the terminals causes Doppler effects. We have seen how the Doppler shift and the angle of arrival of a given echo are interrelated, and that there is a limit to the maximum rate of change possible if it is only MS that is moving. Throughout, we have assumed that multipath gives rise to a spatial standing wave, sensed by the MS antenna as it moves. We have also shown a possible way of generating time variations when MS is stationary. Finally, we briefly introduced the case where both terminals and even the scatterers are moving. In coming chapters, we will present other multipath effects.

References

[1] S.R. Saunders. *Antennas and Propagation for Wireless Communication Systems*. John Wiley & Sons, Ltd, Chichester, UK, 1999.

[2] W.C.Y. Lee. *Mobile Communications Design Fundamentals*. Wiley Series in Telecommunications and Signal Processing. John Wiley & Sons, Ltd, Chichester, UK, 1993.

[3] J.M. Hernando & F. Pérez-Fontán. *An Introduction to Mobile Communications Engineering*. Artech House, 1999.

Software Supplied

In this section, we provide a list of functions and scripts, developed in MATLAB®, implementing the various projects mentioned in this chapter. They are the following:

```
project411            RayleighCDF
project412            fCDF
project413
project421
project422
project431
project432
project433
project434
project44
project45
project46
project471
project472
```

5

Multipath: Narrowband Channel

5.1 Introduction

Now that some of the basic aspects about the multipath channel have been discussed in the previous chapter, here other relevant issues are dealt with. Different alternatives to the multiple point-scatterer model used in Chapter 4 are also presented. In the previous chapter, we did not pay sufficient attention to the amplitudes of the various rays. Here, this point is considered with care so that, when a given *normalization criterion* is used, the simulated results will produce accurate levels, so that actual absolute values can be obtained after putting back the normalization value.

This is very important in link budgets since, at the end of the day, we will be interested in assessing the statistics of the *carrier-to-noise ratio* (CNR) or the *energy per bit-to-noise spectral density ratio* (E_b/N_0) for calculating system performance measures such as the *bit error rate* (BER), *word error rate* (WER), *packet error rate* (PER), *outage* duration, and other relevant quality parameters. Similarly, absolute interference levels and their variability, must be properly characterized so that the calculated *carrier-to-interference ratio* (CIR) statistics are correct.

Here, signal normalization will be performed with respect to the *free space* (FS) level. Thus, a voltage or field strength of one will represent the normalization value. In logarithmic units, this level will correspond to zero dB. Of course, other normalization criteria may be chosen, e.g., with respect to the local average power.

5.2 Projects

Here, we present a series of projects trying to reproduce the behavior of multipath channels through the generation of time series, and their ensuing statistical analyses to gain more insight into this phenomenon. To achieve this goal, we first continue with the multiple point-scatterer model, and reproduce the results obtained by Clarke [1] many years ago. We reproduce both the Rayleigh and Rice conditions. Then, other alternatives to simulating the channel are provided: one based on an array of sinusoidal, low-frequency generators [2]; and another based on using two filtered Gaussian noise generators in quadrature [2]. To finalize

Modeling the Wireless Propagation Channel   F. Pérez Fontán and P. Mariño Espiñeira
© 2008 John Wiley & Sons, Ltd

the chapter, we investigate the spatial properties of the channel (later, in Chapter 10, this investigation is further extended), and look into some diversity and combining concepts. The proposed simulations are not the only ones possible, but should provide the reader with a reasonably good understanding of the narrowband fading phenomenon.

Project 5.1: The Clarke Model

In this project, we will reproduce the so-called *Clarke model* [1] using the multiple point-scatterer approach (`project511`) already employed in Chapter 4. Clarke assumed that the transmitter uses vertical polarization, and that the received vertically polarized field is made up of N plane waves arriving at the antenna with a uniform azimuth distribution, where their phases are arbitrary (uniformly distributed), and their amplitudes are identical.

In `project511` we use signals normalized with respect to the *free space* (FS) level. Thus, if $r(t)$ or $r(x)$ represents the normalized complex envelope at a given point in time, t, or location, x. The normalized voltage, received under FS conditions, will be set such that $|r(t)| = |r(x)| = 1$ or, in dB, $20 \log |r(t)| = 20 \log |r(x)| = 0$ dB.

The actual received power and voltage are linked through, $p = v^2/(2R)$ where R is the load resistance. We normalize with respect to the FS power, i.e.,

$$p' = \frac{p}{p_{FS}} = \frac{\dfrac{v^2}{2R}}{\dfrac{v_{FS}^2}{2R}} = \frac{v^2}{v_{FS}^2} = r^2 \tag{5.1}$$

where p' is the normalized power and r the normalized voltage.

If our normalized voltage is, for example, $r = 0.3$, this is equivalent to a level in dB of $20 \log(0.3) = -10.46$ dB. This means that the associated instantaneous received power will be -10.46 dB below that of FS conditions. Supposing, for example, a transmitted EIRP of 30 dBm. The FS loss at 1 km for a carrier frequency of 2 GHz is

$$L_{FS}(\text{dB}) = 10 \log \left(\frac{4\pi d}{\lambda} \right)^2 = 32.4 + 20 \log(1) + 20 \log(2000) = 98.4 \,\text{dB} \tag{5.2}$$

Then, taking a $G_r = 0$ dBi (dB with respect to the *isotropic antenna*) MS antenna gain, the received power is -68.4 dBm for FS conditions. A fade like the one discussed above (-10.46 dB) will correspond to an instantaneous received power of -78.86 dBm.

We are interested in characterizing the variations of r. We have seen in Chapter 4 that, under multipath propagation conditions such as those of the Clarke model, with the direct signal totally blocked, Rayleigh distributed variations will take place. We know (Chapter 1) how its pdf can be expressed as a function of its parameters: mode, mean, etc.

We can think of characterizing the normalized amplitude variations in terms of the rms value of the signal variations, this value, put in terms of the mode, σ, is given by $\sqrt{2}\sigma$. The normalized average power, $\bar{p}'$, fulfills the expression $\bar{p}' = r_{rms}^2$. On the other hand, we know from Chapter 1 that $r_{rms}^2 = 2\sigma^2$. Thus, we can link these three parameters, that is,

$$\bar{p}' = r_{rms}^2 = 2\sigma^2 \tag{5.3}$$

Usually, given that the direct signal is blocked, and that there is only multipath, the average received power will be several dB below that corresponding to FS. Let us assume that there is a constant *excess loss* (Chapter 1) of 20 dB over a short stretch of the MS route: some tens of wavelengths (small area). We need to link the *normalized average power* for this small area to the magnitudes of the N point-scatterer contributions, a_i. For simplicity, and as in Clarke's model, we can assume that all contributions have the same amplitude, a. We will come back to the issue of the scatterer magnitude in Chapter 7, where we will study its distance dependence.

For example, if we want to simulate the received signal over a small area, with the average power being -20 dB, we will require to adequately select the ray magnitudes as a function of the number of rays, N, we want to use. Calculating the rms squared value first and then the modal value of the Rayleigh distribution for the given normalized average power level, we get

$$\bar{P}' = -20\,\text{dB} = 10\log(\bar{p}') \Rightarrow \bar{p}' = 0.01 = 2\sigma^2 \Rightarrow \sigma = 0.0707 \qquad (5.4)$$

This normalized average power should be equal to the *power sum* of the various multipath amplitudes, i.e.,

$$\bar{p}' = 2\sigma^2 = \sum_{i=1}^{N} a_i^2 = Na^2 \qquad (5.5)$$

This, of course, is an average value but the actual time series will show a wide range of larger and smaller instantaneous powers/voltages.

Depending on the parameter we want to simulate, we will need more or less rays. If for example, as in Project 4.4, we chose $N = 13$ rays of identical amplitudes, a, this value must be

$$0.01 = Na^2 \Rightarrow a = 0.0277 \qquad (5.6)$$

If we used 10 times more rays, i.e., 130, the new amplitude would be $a = 0.0088$.

Associated to the Rayleigh distributed amplitudes, we can work out the distribution of the normalized power, $p' = r^2$. Performing this change of variable, we get

$$f(p') = \frac{1}{2\sigma^2}\exp\left(-\frac{p'}{2\sigma^2}\right) = \frac{1}{\bar{p}'}\exp\left(-\frac{p'}{\bar{p}'}\right) \quad \text{for} \quad p' \geq 0 \qquad (5.7)$$

which is an *exponential distribution* of mean $\bar{p}' = 2\sigma^2$. In exponential distributions, the mean is equal to the standard deviation, thus, the value of the latter parameter is also $2\sigma^2$.

The simulation in `project511` is basically a repetition of the one carried out in Project 4.4. Clarke assumed that typical multipath conditions imply a uniform distribution of scatterers about MS, all on the same horizontal plane of the MS antenna. The effects of 3D distributions of scatterers have been studied by many researchers, and the basic results are summarized in [3].

To simulate a uniform angle of arrival distribution of point scatterers, we can set them on a circle about MS as illustrated in Figure 5.1. The simulator settings used are as follows: BS is located 1 km from the origin; the scatterers are located on a circle of

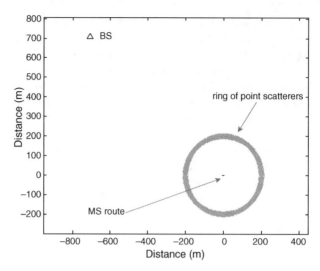

Figure 5.1   Point scatterers on a circle about MS to simulate a uniform angle of arrival distribution

radius 200 m centered at the origin. Their positions (angles in this case) are drawn randomly using MATLAB® (MATLAB® is a registered trademark of The MathWorks, Inc.) function rand which generates uniformly distributed samples: rand(1000,1)*360. The number of scatterers has been set to 1000. For calculating the Doppler spectra, shown below, the number of FFT points was set to 1024. The route sampling rate was of 16 points per wavelength. Note that the MS route is along the x-axis and contains the origin.

In addition to the results relative to the magnitude of the received complex envelope, which we already know should match a Rayleigh distribution, a number of results are provided in the literature for the Clarke model [1][2]. Maybe the most well known is that referring to its U-shaped Doppler spectrum following the expression,

$$S(f) = \begin{cases} k[1 - (f/f_m)^2]^{-1/2} & |f| \leq f_m = V/\lambda \\ 0 & \text{elsewhere} \end{cases} \tag{5.8}$$

where $S(f)$ is the spectral density and k is a constant. This spectrum becomes infinite at $\pm f_m$, the maximum Doppler shift. As discussed in [3], more realistic Doppler spectra, with limited levels in the neighborhood of $\pm f_m$, can be obtained if the angles of arrival of the scattered contributions are not all on the horizontal plane of the MS antenna, but show some spreading also in elevation.

Figure 5.2(a) shows the simulated received signal envelope expressed in dB, now varying well below the LOS level (0 dB), more specifically about -20 dB as to be expected. Its distribution matches fairly well a Rayleigh distribution as shown in Figure 5.2(b). The phase is shown in Figure 5.3, and in Figure 5.4 the histogram of the phase is plotted, clearly indicating that it is uniformly distributed. Figure 5.5(a) shows the Doppler spectrum for the complex envelope, where the predicted U-shaped spectrum can clearly be observed. Figure 5.5(b) shows the spectrum of the magnitude of the complex envelope, $|r|$, with a limit at $2f_m$, as predicted in the theory [1].

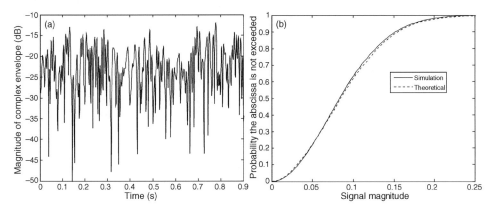

Figure 5.2 (a) Simulated normalized signal magnitude in dB. (b) Cumulative distribution of simulated signal and theoretical Rayleigh CDF

Some verifications as to the goodness of the model implemented can be performed. In [1] it is indicated that the autocorrelation of the complex envelope follows the expression,

$$\rho(\Delta t) = J_0(2\pi f_m \Delta t) \quad \text{or} \quad \rho(\Delta x) = J_0\left(\frac{2\pi}{\lambda}\Delta x\right) \tag{5.9}$$

where Δt is the time lag between samples, and Δx is the corresponding distance lag, where a constant MS speed, V, is assumed. J_0 is the Bessel function of first kind and zero order, available in MATLAB® as function `besselj`. Figure 5.6 shows the simulated and theoretical autocorrelations. The theory also indicates that the cross-correlation between the in-phase, I, and quadrature, Q, rails of the complex envelope, i.e., the real and imaginary parts, is zero. Also, that both I and Q shall be Gaussian distributed with mean zero and standard deviation, σ, equal to the mode of the associated Rayleigh for the amplitude. These tests are left for the reader to perform.

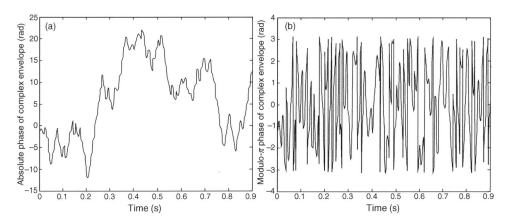

Figure 5.3 (a) Absolute phase. (b) Modulo-π phase

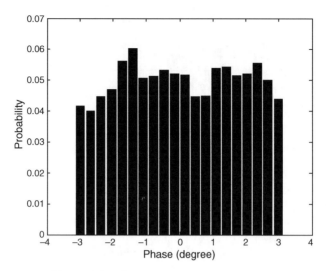

Figure 5.4   Histogram of the simulated phase

Second-order statistics. Next, other parameters frequently found in the literature are presented and simulation results using `project511` are shown. One such parameter is the *level crossing rate*, lcr, and another is the *average fade duration*, afd. These parameters are often called *second-order statistics* as they depend on the mobile speed. The level crossing rate is defined as the average number of times the signal crosses a given threshold, R, within a given observation period, T, with either a positive or negative slope. This is illustrated in Figure 5.7 for a relative level of -20 dB/LOS. On the other hand, the average fade duration is the ratio between the total time the received signal is below a reference level, R, and the total number of fades. These two parameters are of interest since they can help in the selection of the most suitable error protection coding scheme and interleaving algorithm. The afd helps determine the most likely number of bits that may be lost during a fade.

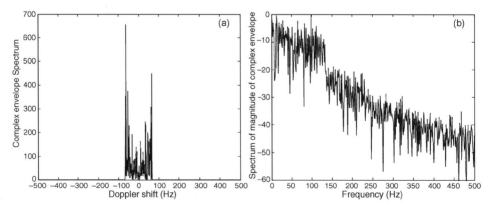

Figure 5.5   (a) Doppler spectrum of synthesized complex time series. (b) Spectrum of magnitude of complex envelope

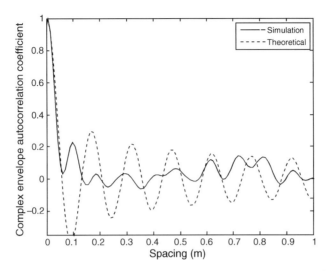

Figure 5.6   Autocorrelation of the simulated complex envelope and theoretical model

In `project511`, MATLAB® function `find` is used for locating the vector elements (e.g. in a time series) above or below a given threshold.

In the literature [3], theoretical models can be found for these two parameters. Here, we show results obtained using the same basic assumptions as in Clarke's model. The level crossing rate, N_R, is defined in a rigorous way as follows,

$$N_R = \int_0^\infty \dot{r} f (R, \dot{r}) d\dot{r} = \sqrt{2\pi} f_{\mathrm{m}} \rho \exp(-\rho^2) \qquad (5.10)$$

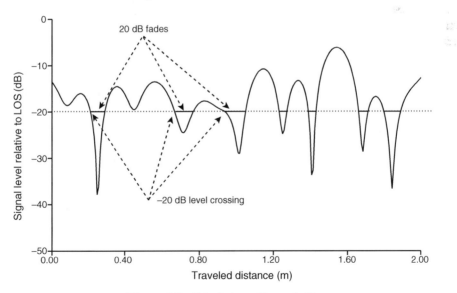

Figure 5.7   Calculation of lcr and afd

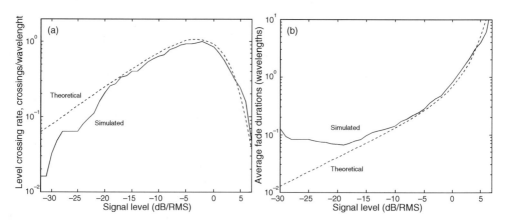

Figure 5.8   (a) Simulated and theoretical lcr. (b) Simulated and theoretical afd

where $\dot{r}$ is the time derivative (slope) of the magnitude of the complex envelope, and $f(R, \dot{r})$ is the joint probability density function of the magnitude particularized at $r = R$ and its derivative. The above equation also provides a model where its dependence on the mobile speed comes in through parameter f_{m}, the maximum Doppler shift. In the above formula, the amplitude is normalized with respect to the rms value, i.e., $\rho = R/(\sigma\sqrt{2})$.

From the observation of the curve corresponding to the lcr formula, shown in Figure 5.8(a), it can be pointed out that there are few crossings both at the high and low levels, while the maximum number occurs at $\rho = 1/\sqrt{2}$, i.e., 3 dB below the rms level. Figure 5.8(a) also shows the computed lcr from the simulated time series. The lcr values are normalized and given in terms of crossings per wavelength, i.e., parameter N_R/f_{m} has been plotted. Very good agreement between simulated results and the model can be observed except for the very low levels which are lost in the sampling process.

The afd can be calculated by counting the total time below a given level divided by the total number of fades (crossings), i.e.,

$$L_R = \mathrm{Prob}(r \leq R)/N_R \tag{5.11}$$

where $\mathrm{Prob}(r \leq R) = \sum_i t_i/T$, with T the observation time, is the CDF which, for a Rayleigh distribution is given by

$$\mathrm{Prob}(r \leq R) = \int_0^R f(r)dr = 1 - \exp(-\rho^2) \tag{5.12}$$

Alternatively, the afd can be computed by keeping track of all individual fade durations, and computing their mean value, i.e.,

$$\mathrm{adf}(R) = (D_1 + D_2 + \cdots + D_N)/N \tag{5.13}$$

This information, i.e., the durations of individual fades, can be used for calculating the *distribution of fade durations* – not only their average. Calculating and plotting the distribution of fade durations for a given reference level is a task left for the reader to implement.

A model for the afd is provided in [3],

$$L_R = \frac{\exp(\rho^2) - 1}{\sqrt{2\pi} f_{\mathrm{m}} \rho} \qquad (5.14)$$

Figure 5.8(b) shows both the theoretical model and the simulated results for the afd. As in the case of the lcr, the durations are normalized and presented in wavelengths, i.e., $L_R f_{\mathrm{m}}$ have been plotted. Very good agreement can be observed between the simulated and the theoretical results except for the very low signal levels, which are lost in the sampling process.

A justification for the very high sampling frequency or fraction, F, of the wavelength needed can be obtained through simple calculations [3] of the afd for different signal levels. Thus, in Table 5.1 the afd and lcr values for different levels are provided. For example, from the table, in order to detect about 50% of the fades 30 dB below the rms level, the signal must be sampled every 0.0126λ, which means a fraction, F, of the wavelength of approximately 79.

Random FM. So far, we have seen that the mobile channel gives rise to random amplitude and phase variations, together with Doppler shifts. One further effect caused by the channel is a random frequency modulation, *random FM*, which is more marked at the deep fades. This can be considered as an additional noise source affecting the transmitted signal, especially if a frequency-sensitive detector is used. The random FM caused by the channel can be calculated by taking the derivative of the received signal phase, i.e.,

$$\dot{\theta} = \frac{d\theta}{dt} = \frac{d}{dt}\left[\tan^{-1}\frac{Q(t)}{I(t)}\right] \qquad (5.15)$$

In project511 MATLAB® function diff is used for performing the discrete differentiation of the phase (angle). The results must then be divided by the sampling interval. Figure 5.9 illustrates the simulated received signal amplitude, its phase and random FM. Note how the deep fades give rise to sharp phase changes and strong random FM peaks.

Doppler and fade rate. Several interesting simulations can be carried out using the same simulator. For example, we can look into the close relationship between the Doppler spread and the fade rate, already discussed in Project 4.3.

Table 5.1 Average fade length and crossing rate for fades relative to the rms level [3]

Level wrt rms level (dB/rms)	afd (λ)	lcr (λ^{-1})
0	0.6855	0.9221
−10	0.1327	0.7172
−20	0.0401	0.2482
−30	0.0126	0.0792
−40	0.0040	0.0251
−50	0.0013	0.0079

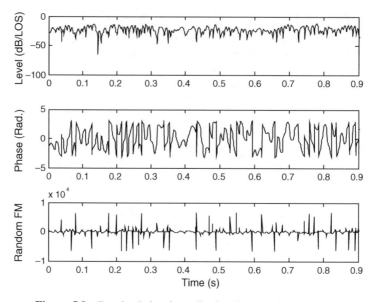

Figure 5.9   Received signal amplitude, phase, and random FM

Figure 5.10 [1] illustrates, both geometrically and in terms of Clarke's Doppler spectrum, the basics of simulators `project512`, `project513` and `project514`. What we want here is to show the clear relationship between the width of the Doppler spectrum and the rate of change of the fades, i.e., how fast they are. Basically, the Doppler spectrum is the representation of the fading process in the frequency domain. As will be shown in Project 5.4, the selectivity and shape of the Doppler spectrum can be used to limit the bandwidth of the fading process. A wider spectrum means faster variations. In Figure 5.10 a directive

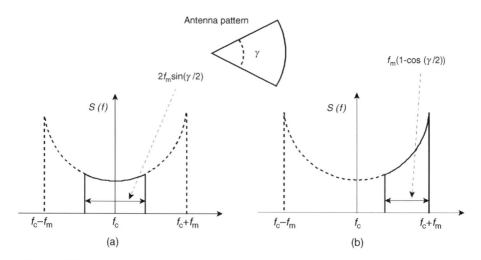

Figure 5.10   Doppler spectrum when an ideal directive antenna is assumed at the MS [1]

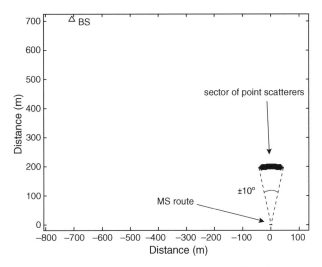

Figure 5.11 Scatterers seen with a directive antenna pointing 90° with respect to the direction of travel and a beamwidth of ±10°

antenna is assumed, which has an ideal, brick-wall pattern with constant unit gain in the angular sector toward which it is pointing, and zero gain elsewhere.

To simulate the effect of the MS antenna directivity, we will modify the simulation scenario so that only those scatterers in the field of vision of the assumed antenna are present. We want to look at the Doppler spectrum and the fade rate. Figures 5.11–5.16 illustrate three cases where ideal directive antennas have been assumed. The first case, project512, corresponds to the situation where the ideal sector antenna is pointing in a direction 90° with respect to the direction of travel (x-axis), and has a beamwidth of ±10° (Figure 5.11). The reader is reminded that the mobile route passes over the origin of coordinates. We can see in Figure 5.12(a) the corresponding signal variations and in Figure 5.12(b) the associated Doppler spectrum. The maximum shift in the onmi-azimuth

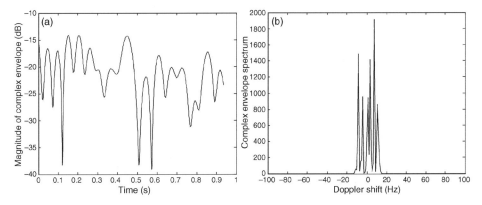

Figure 5.12 (a) Simulated signal amplitude variations and (b) simulated Doppler spectrum for the geometry in Figure 5.11

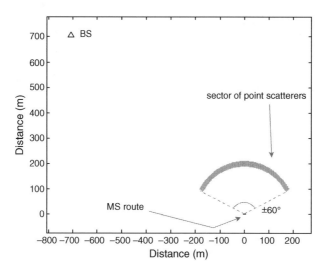

Figure 5.13 Scatterers seen with a directive antenna pointing 90° from the direction of travel and a beamwidth of ±60°

case should be 66.67 Hz, here it is much smaller. To increase the fade rate we can widen the sector of the MS antenna radiation pattern, `project513`. In this case, a width of ±60° is assumed (Figure 5.13). We can clearly see how the fade rate increases (Figure 5.14(a)) at the same time as the Doppler spectrum widens (Figure 5.14(b)). One final case is that of `project514` where a ±10° wide antenna is looking in the direction of travel (Figure 5.15). In this case, the rate of change is much slower (Figure 5.16(a)) and the Doppler spectrum much narrower (Figure 5.16(b)). In Figure 5.10, the repercussions of the antenna beamwidth and pointing angle on the Doppler spectrum are shown schematically; equations are also provided in this figure.

Some or most of the previous results are normally obtained in books dealing with the mobile channel through involved statistical-mathematical formulations. The theoretical

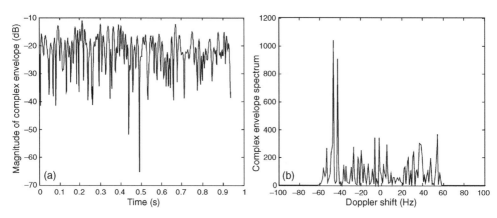

Figure 5.14 (a) Simulated signal amplitude variations and (b) simulated Doppler spectrum for the geometry in Figure 5.13

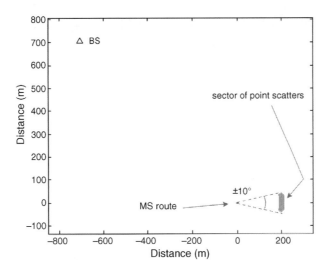

Figure 5.15 Scatterers seen with a directive antenna pointing in the direction of travel and a beamwidth of $\pm 10°$

results fit quite well with the simulations performed here. This may allow the reader to become familiar with the multipath phenomenon, and some of its basic results, in a more intuitive, simulation-based way.

Project 5.2: The Rice Channel

The Rayleigh model for signal fading is really a worst-case scenario. It postulates that the direct signal is totally blocked. Fading conditions, when the direct signal is unblocked or slightly attenuated (shadowed), can be quite well described by using the Rice distribution. In project52, we make the same assumptions as in project511, the only change is that a

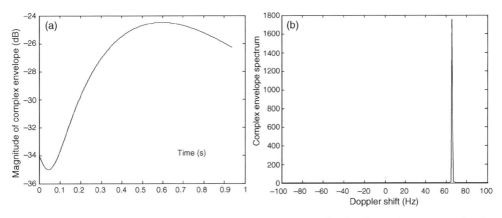

Figure 5.16 (a) Simulated signal amplitude variations and (b) simulated Doppler spectrum for the geometry in Figure 5.15

new contribution, i.e., the direct ray between BS and MS is present. Its amplitude, a, can take up values ranging from one, meaning unblocked, LOS conditions, to zero, meaning a totally blocked direct component, which is equivalent to the Rayleigh case. The Rice pdf is given by equation

$$p(r) = \frac{r}{\sigma^2} \exp\left(-\frac{r^2 + a^2}{2\sigma^2}\right) I_0\left(\frac{ar}{\sigma^2}\right) \quad r \geq 0 \tag{5.16}$$

Again, r actually represents the magnitude of the complex envelope, i.e., $|r|$. I_0 is the modified Bessel function of first kind and zero order, which is available in MATLAB®- (besseli). A commonly used parameter is the so called *Rice k-factor*, defined as

$$k = a^2/2\sigma^2 \quad \text{and, in dB,} \quad K(\text{dB}) = 10\log(k) \tag{5.17}$$

also known as the *carrier-to-multipath ratio* (C/M). Parameter a is the amplitude of the direct signal and σ is the mode of the associated Rayleigh distribution resulting when a is equal to zero, with $2\sigma^2$ being the normalized average power of the multipath component. Note in Figure 5.17 the evolution of the Rice pdf toward a Rayleigh distribution as the direct signal decreases.

Figure 5.18 shows the simulated scenario, the same results as in project511 are shown in Figures 5.19 and 5.20, corresponding to a C/M value of approximately 20 dB, where the direct signal amplitude is one and the average multipath power is again $-20\,\text{dB}(2\sigma^2)$ as in preceding simulations. Note the drastic change in received signal levels. Now the oscillations are about the zero dB level, while before they were close to $-20\,\text{dB}$. The dynamic range of the oscillations has also changed, now the fades are less deep than in the Rayleigh case, showing how, when the direct signal is present (for example at a crossroads with direct view of BS), the channel is much milder, posing less problems for the transmitted signal.

Figure 5.19(a) shows the signal level variations in dB while Figure 5.19(b) shows the series CDF together with the theoretical Rice distribution function, the fit being quite good. Figure 5.20(a) illustrates the variations in the phase. As MS drives toward BS, the phase increases. Superposed on the dominant linear phase variations due to the direct signal, very small phase variations due to multipath can be observed when zooming in on the MATLAB®

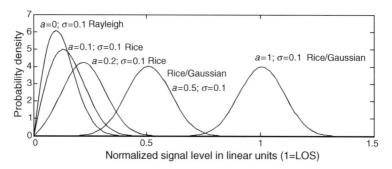

Figure 5.17 Evolution of the Rice distribution pdf toward a Rayleigh distribution as the direct signal decreases

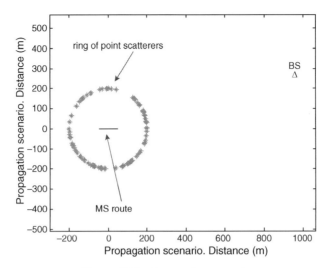

Figure 5.18 Propagation scenario

figure. The same phenomenon can also be viewed on the Doppler spectrum (Figure 5.20(b)) where, dominating over the diffuse multipath U-shaped spectrum, is a clear positive Doppler line due to the direct signal.

Project 5.3: Rayleigh Channel Simulator using Sinusoidal Generators

The multiple point-scatterer model can be implemented both in software and in hardware assuming a limited number of sinusoidal, low-frequency generators. The schematic diagram of this generator is shown in Figure 5.21. The basic assumption that needs to be made is equivalent to that of having parallel rays from each scatterer to each route sampling point.

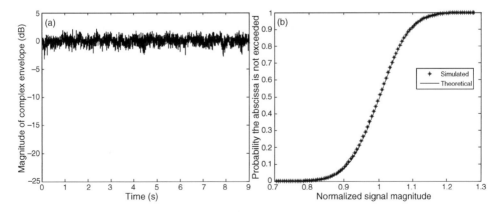

Figure 5.19 (a) Simulated signal level variations in dB. (b) Simulated series CDF and Rice theoretical CDF

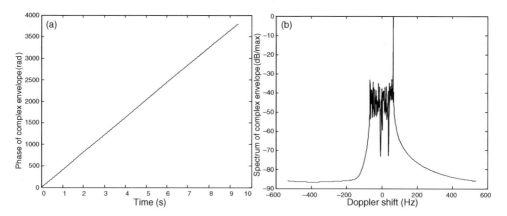

Figure 5.20   (a) Absolute phase variations. (b) Doppler spectrum

In this case, the relationship

$$f_{Di} = \frac{V}{\lambda} \cos(\theta_i) = f_m \cos(\theta_i) \tag{5.18}$$

is fulfilled (Chapter 4) for the Doppler shift originating from scatterer i with an angle θ_i with respect to the mobile route. This assumption can be sustained if the simulated route

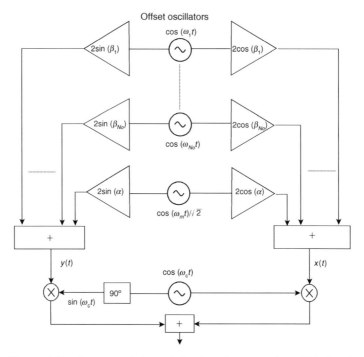

Figure 5.21   Structure of channel simulator based on sinusoidal signals

is short enough, i.e., a small area. Then, the received complex envelope can be put in the form

$$r(t) = \sum_{i}^{N} \exp\{j[2\pi f_m \cos(\theta_i)t + \phi_i]\} \tag{5.19}$$

where a constant amplitude, as in Project 5.1, has been assumed for all multipath contributions, and where ϕ_i is the initial phase of contribution i. This expression can be rearranged by assuming that, for each positive Doppler spectral line, there exists an associated negative line, thus

$$r(t) = \sum_{i}^{N/2} \left(\exp\{j[2\pi f_m \cos(\theta_i)t + \phi_i]\} + \exp\{-j[2\pi f_m \cos(\theta_i)t - \phi_i]\} \right)$$

$$\tag{5.20}$$

$$= 2 \sum_{i}^{N/2} \cos(\phi_i) \cos[2\pi f_m \cos(\theta_i)t] + j2 \sum_{i}^{N/2} \sin(\phi_i) \cos[2\pi f_m \cos(\theta_i)t]$$

where it is clearly shown how the complex envelope is made up of in-phase and quadrature components, consisting of sinusoidal generators (cosines), followed by amplifiers (multiplication factors).

In [2] recommendations are given as to what settings are best for implementing this approach. The expression implemented in `project53` separates from the group of sinusoids the term corresponding to the maximum Doppler f_m. Thus, the overall in-phase and quadrature components are

$$I(t) = 2 \sum_{n=1}^{N_0} \cos(\beta_n) \cos(2\pi f_n t) + \sqrt{2} \cos(\alpha) \cos(2\pi f_m t)$$

$$\tag{5.21}$$

$$Q(t) = 2 \sum_{n=1}^{N_0} \sin(\beta_n) \cos(2\pi f_n t) + \sqrt{2} \sin(\alpha) \cos(2\pi f_m t)$$

We can implement the simulator using $N_0 + 1$ sinusoidal, low frequency ($\leq f_m$) generators followed by amplifiers, different for the in-phase and quadrature rails (Figure 5.21). The frequencies of the generators shall be

$$f_m \cos(2\pi n/N) \quad n = 1, 2, \ldots N_0 \tag{5.22}$$

plus one at frequency f_m, where $N_0 = 0.5(0.5N - 1)$. The gains must be $2\cos(\beta_n)$ and $2\sin(\beta_n)$. For the generator at f_m, the gain must be $\sqrt{2}$ and $\beta_n = \pi n/N_0$.

The block diagram in Figure 5.21 also shows the RF signal – not only the complex envelope. This was one of the first channel emulators built before digital signal processors (DSPs) were widely available, since it required a very simple analog implementation.

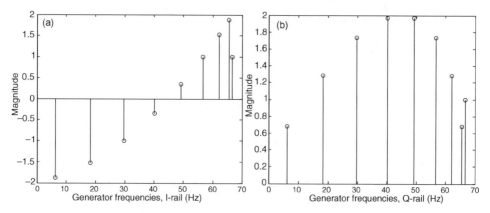

Figure 5.22 (a) Sinusoidal generators for the *I* rail. (b) Sinusoidal generators for the *Q* rail

When an unmodulated RF carrier of frequency f_c is transmitted, the received RF signal, $s(t)$, is given by

$$s(t) = \mathrm{Re}[r(t)\exp(\mathrm{j}2\pi f_c t)] = \mathrm{Re}\{[I(t) + \mathrm{j}Q(t)][\cos(2\pi f_c t) + \mathrm{j}\sin(2\pi f_c t)]\}$$
$$= I(t)\cos(2\pi f_c t) - Q(t)\sin(2\pi f_c t) \tag{5.23}$$

where the linkage between the $I(t)$ and $Q(t)$ elements of the complex envelope, $r(t)$, and the carrier term is shown. The example given in [2] assumes $N_0 = 8$, and the gains associated to the oscillator are controlled by parameter $\alpha = \pi/4$ (Figure 5.22).

Figure 5.23(a) presents the resulting amplitude time series, and Figure 5.23(b) the sample CDF together with the corresponding theoretical Rayleigh CDF, showing an excellent agreement. Figure 5.24 plots the phase and Figure 5.25(a) its histogram, showing that it is uniformly distributed. Finally, Figure 5.25(b) shows the resulting Doppler spectrum made up of discrete lines, as to be expected.

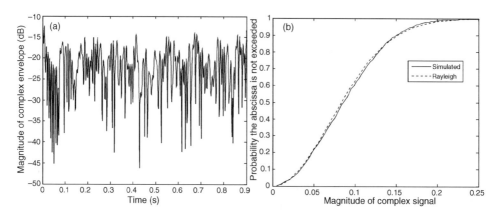

Figure 5.23 (a) Simulated time series in logarithmic units. (b) Sample CDF and theoretical Rayleigh CDF

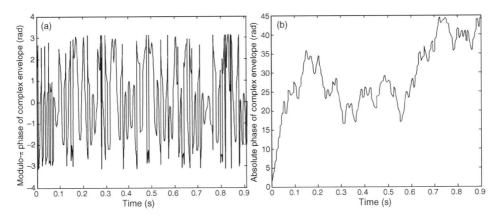

Figure 5.24   (a) Modulo-π phase. (b) Absolute phase

Some further tests can be carried out for verifying the goodness of the simulator. As suggested in Project 5.1, we can perform several verifications including that the autocorrelation fits the expected behavior, that the I and Q rails are Gaussian with standard deviation equal to the modal value of the Rayleigh distribution of the amplitude, and the amplitudes of the two rails are uncorrelated. These results are shown in Figures 5.26 and 5.27.

Project 5.4: Simulator using Two Filtered Gaussians in Quadrature. Butterworth Response

In this project, we are going to simulate, in parallel, both a Rayleigh and a Rice time series, by means of two Gaussian random number generators plus a filter. In previous projects, we have seen how the Doppler spectrum is related to the fade rate, and how the multipath phenomenon can be reproduced by using a more or less large number of point scatterers. We have also seen how each scatterer gives rise to a Doppler line, and how we can perform the simulation using a limited number of sinusoidal generators, equivalent to using discrete scatterers.

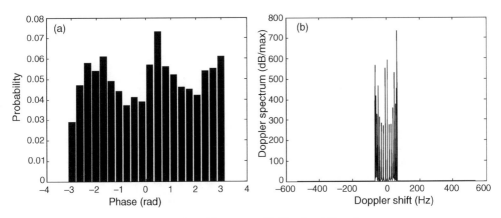

Figure 5.25   (a) Phase histogram. (b) Simulated Doppler spectrum

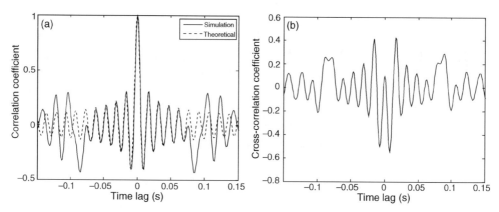

Figure 5.26   (a) Autocorrelation of the complex envelope and theoretical model. (b) Cross-correlation between the I and Q rails

In fact, the multipath channel is actually produced by a continuum of contributions, i.e., an infinite number of scatterers. The central limit theorem predicts that, as the number of random variables increases, its sum tends to a Gaussian distribution. Thus, in `project54` we simulate the mobile channel by using two Gaussian random generators in quadrature: real part and imaginary part of the complex envelope (Figure 5.28). These two generators will produce a flat spectrum limited by $\pm f_s/2$, where f_s is the sampling frequency. As we have learned, the fading phenomenon is band-limited by the width of the Doppler spectrum. We will hence, filter the in-phase and quadrature rails with a Doppler filter to adequately shape the fades.

For creating the two Gaussian time series, MATLAB® function `randn` can be used, which produces a series of uncorrelated samples corresponding to a distribution with zero mean and unit standard deviation. For producing a Gaussian series with mean, a, and

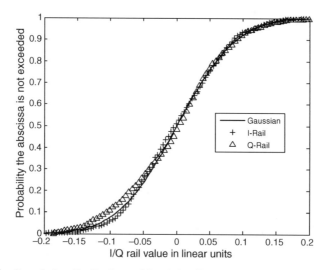

Figure 5.27   Cumulative distributions of I and Q rails and theoretical Gaussian distribution

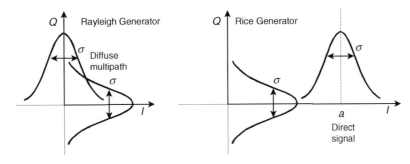

Figure 5.28   Gaussian generators in quadrature for simulating Rayleigh and Rice fades

standard deviation, σ, the series out of `randn` must be multiplied by σ and then, add the mean, a, e.g.,

```
x=randn(1000,1)*sigma+a
```

The average value, a, represents the direct signal and the standard deviations, σ, the multipath. The complex envelope in both the Rayleigh and Rice cases is given by the complex sum,

$$r = I + jQ \qquad\qquad (5.24)$$

which, in MATLAB® code, becomes

```
rRayleigh=randn(Nsamples,1)*sigma+j*randn (Nsamples,1)*sigma;
```

for the *Rayleigh case*, and

```
rRice=(a+randn(Nsamples,1)*sigma)+j*randn(Nsamples,1)*sigma;
```

for the *Rice case*, where we have placed the direct component on the in-phase rail. The direct signal phase is kept constant throughout the simulation. This is equivalent to having an angle of arrival perpendicular to the mobile route. In a more involved simulator, the phase could be made to rotate, for example, at a constant rate, which would mean that the angle of arrival would be other than 90°. We further discuss this in Project 6.3. In any event, the maximum possible Doppler shift should never exceed V/λ.

The following settings have been used in the simulations: $\sigma = 0.12$ and $a = 1$, the sampling frequency, f_s, has been set to 120 Hz. In the first part of this project, we do not introduce a band-limiting (Doppler) filter. This means that the Gaussian sources will have a flat spectrum from $-f_s/2$ to $+f_s/2$. The resulting fades will be too fast. Figure 5.29 illustrates the resulting Rayleigh and Rice time series in dB. The time axis was created using a time step, $t_s = 1/f_s$. Figure 5.30 shows the CDFs of the simulated series and their corresponding theoretical CDFs, the agreement being quite good.

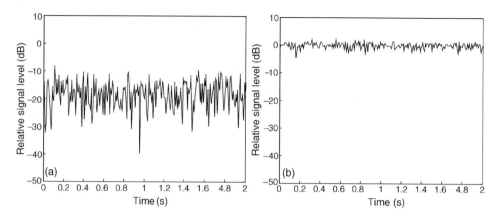

Figure 5.29   (a) Rayleigh series. (b) Rice series

So far, we have been able to reproduce the ordinate part of the time series, i.e., the signal level variations. However, we do not yet have control over the abscissas, i.e., the rate of change or bandwidth of the variations. Figures 5.31(a) and (b) show the spectra of the generated complex time series. As is apparent from the figures, they both show a flat spectra except that, in the Ricean case, a strong delta is present at the origin, corresponding to the direct signal, much stronger than the diffuse multipath component.

Our next step is controlling the bandwidth of the simulated series. This can be achieved through filtering. Figure 5.32 schematically illustrates how this is implemented in the Rayleigh and Rice cases.

In `project54` we are going to assume a Doppler filter different from the one proposed by Clarke. In this case, a *Butterworth filter* will be used. This is a quite common assumption in land mobile satellite channels [4], which, geometrically, implies that the contributions from both ends of the street produce less multipath while most contributions arrive from the sides: the building faces. In Chapter 6 we will generate

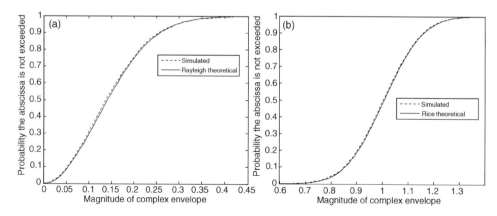

Figure 5.30   Simulated and theoretical CDFs for (a) Rayleigh series and (b) Rice series

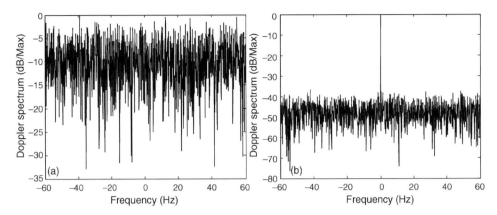

Figure 5.31 (a) Doppler spectrum of generated Rayleigh complex envelope (before Doppler filtering). (b) Doppler spectrum of generated Rice complex envelope (before Doppler filtering)

several time series using a filter approximating Clarke's Doppler spectrum. The Butterworth spectrum follows the expression,

$$|H_{\text{Butt}}(f)|^2 = \frac{A}{1 + (f/f_c)^{2k}} \tag{5.25}$$

where f_c is the cutoff frequency and k is the order of the filter. This filter is available in MATLAB® through functions `buttord`, `butter` and `filter`. It is important to bear in mind that MATLAB®'s implementation changes the standard deviation if a zero-mean Gaussian signal is passed through this filter. Thus, a multiplicative factor, proportional to the inverse of the square root of the filter's energy needs to be introduced at its output to compensate for this effect. MATLAB® functions `freqz` and `impz` can be of help in this and other tasks, such as viewing the channel time and frequency responses.

For simulating the Butterworth filter, we have used the following settings as inputs to MATLAB® function `buttord`: Wp=0.1, Ws=0.3, Rp=3, Rs=40. These parameters define the pass-band and the stop-band of the filter. Frequencies are given with respect to half the sampling frequency, and the attenuations are in dB. Figure 5.33 shows the frequency and impulse responses of the Butterworth filter used in the simulations. Figure 5.34 shows the

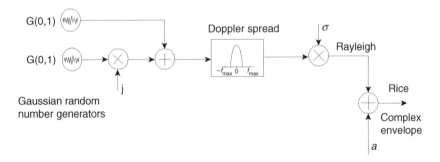

Figure 5.32 Schematic diagram of the Rayleigh/Rice simulator

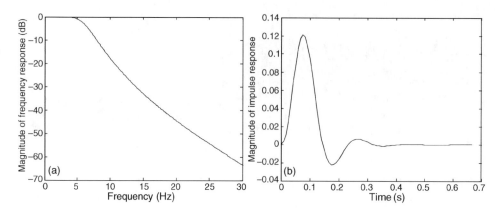

Figure 5.33 Butterworth filter response (a) in the frequency domain and logarithmic units, and (b) in the time domain and linear units

filtered time series. Actually, only the complex Rayleigh series has been filtered. Observe how the variations are now much slower. In the Rice case, a constant, a, has been added to the in-phase rail of the complex filtered Rayleigh (Figure 5.32).

When compared with the unfiltered series, it is clear that they are much slower. It is very important to remember to perform the filtering operation when simulating the channel given that, otherwise, the obtained amplitude and phase variations would be unnaturally fast for a given mobile speed, thus yielding wrong results in terms of error bursts, distribution of outage durations, etc. Figure 5.35 shows the CDFs of the simulated series before and after filtering for both cases. As can be observed, the distributions have not changed after filtering (ordinates of the time series), even though the abscissas have. Figure 5.36 shows their respective Doppler spectra after filtering where the band-limiting effect can clearly be observed.

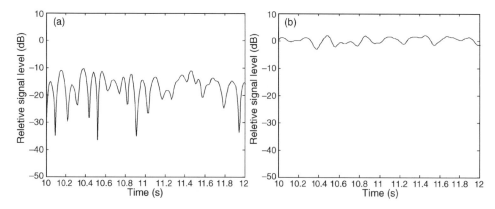

Figure 5.34 (a) Filtered Rayleigh series. (b) Filtered Rice series

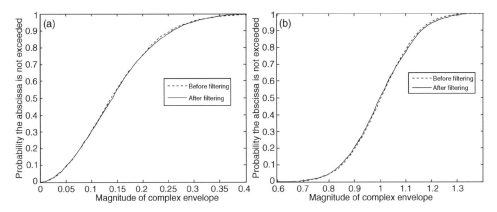

Figure 5.35   Simulated series CDFs for unfiltered and filtered cases. (a) Rayleigh and (b) Rice

Project 5.5: Diversity and Combining at the Mobile Station

In this project, we are going to try and work out what is the distance needed between MS antennas for achieving a reasonably good space diversity gain. MS diversity is rarely implemented due to lack of space, and the added complexity. However, this study allows us to learn more about the nature of the scattering channel. Basically, a diversity scheme consists of two elements, namely, the *diversity part* where at least two, preferably uncorrelated, but normally partially correlated signals must be provided to the second element, the *combiner*. Figure 5.37 illustrates several possible combiners [5]: selective, switched, maximal ratio and equal gain. The diversity branches can be provided in space, time, angle, polarization, etc. [3].

Here, we concentrate on space diversity at MS. Figure 5.38 illustrates the simulation scenario corresponding to a macrocell. The MS receiver is surrounded by point scatterers while BS is not, because it is on an elevated position, clear of local scatterers. We are going to consider that the two antennas providing the two diversity branches are aligned with the MS route. This is mainly for ease of implementation of the simulator.

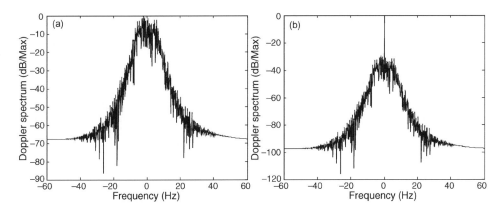

Figure 5.36   Doppler spectra of (a) filtered Rayleigh series and (b) filtered Rice series

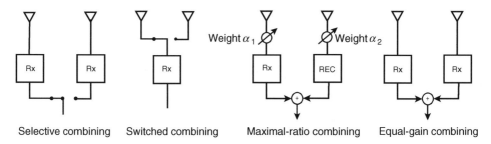

Selective combining   Switched combining   Maximal-ratio combining   Equal-gain combining

Figure 5.37   Combiner types [5]

Our first step is computing the *cross-correlation* between the two signals in each diversity branch. We are interested in finding what separation is necessary for having two weakly correlated signals. Assuming the two antennas are aligned with the mobile route means that the signal received by each of the antennas is the same but slightly offset in time/space. In this case, we are interested in presenting the received signals with the abscissas in distance units: meters or wavelengths.

Thus, the wanted cross-correlation of the signals on the two diversity branches can be calculated as the autocorrelation of the received signal, the different lags representing different antenna separations. Unlike in Project 5.1 where the autocorrelation was computed for the complex envelope, here we concentrate on the autocorrelation of the signal magnitude, $|r|$, since we are interested in compensating the fades.

Figure 5.39(a) illustrates the autocorrelation of the received signal magnitude as a function of separation which, for the antennas aligned with the route, correspond to the

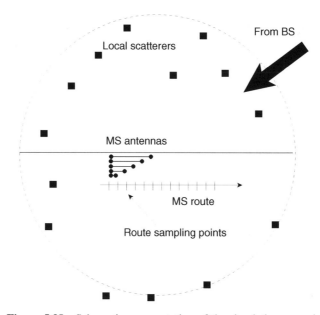

Figure 5.38   Schematic representation of the simulation scenario

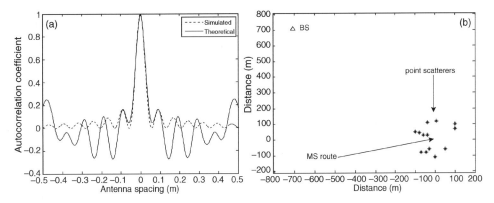

Figure 5.39 (a) Simulated and theoretical autocorrelation function for the magnitude of the complex envelope at MS. (b) Actual simulated scenario. MS is at origin. BS is represented by triangle. Asterisks are point scatterers

cross-correlation for different antenna spacings. Also in the figure, the *theoretical model* [3] which follows the expression

$$\rho_{|r|}(\Delta x) = J_0^2\left(\frac{2\pi}{\lambda}\Delta x\right) \tag{5.26}$$

is presented. It is clear from the figure, that a separation of even less than half the wavelength, already provides acceptably low values of the cross-correlation coefficient. The selection of the antenna spacing should be based on this criterion.

Figure 5.39(b) illustrates the actual scenario simulated in `project55`. In Figures 5.40(a)–(d) show the two diversity signals and the result of its combination based on always switching to the best signal (*selection combining*) for different spacings. These figures illustrate the advantage of introducing space diversity as a countermeasure against multipath fading. It can be observed that, as the spacing increases, the resulting combined signal presents shallower fades. The same results are summarized in Figure 5.41 where the cumulative distributions of the original and the combined signals are shown. It is clear that diversity makes the fading statistics less harsh for reception, thus reducing the total outage times. The durations of fades are also drastically reduced. This is left for the reader to verify.

Finally, Figure 5.42 illustrates two concepts normally used for describing the performance of diversity schemes: the *diversity gain* and the *diversity improvement*. The reader is encouraged to calculate these parameters from the variables in the MATLAB® workspace.

Project 5.6: Diversity and Combining at the Base Station

Even though the reciprocity theorem is valid for the MS–BS link, some characteristics of a scattering medium are not reciprocal. This is the case of the cross-correlation between space diversity antennas at BS. Unlike at MS, macrocell BSs are not normally surrounded by scatterers as they are usually sited at a sufficient height to be clear of nearby obstacles. Thus, the BS antenna will see the scatterers within a much narrower look-angle/sector, while MS is surrounded by scatterers as illustrated in Figure 5.43. Also in the figure, the basic elements in the developed simulators (`project561`, `project562` and `project563`) are depicted.

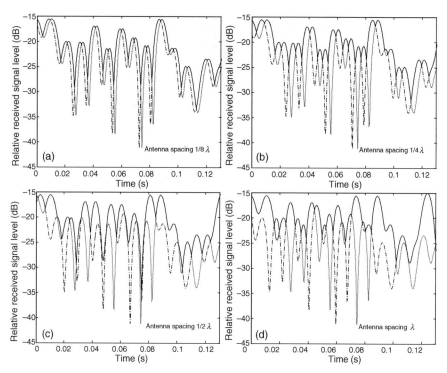

Figure 5.40 Received signals on both diversity branches and selection combined signal for an antenna spacing equal to (a) ($\lambda/8$, (b) $\lambda/4$, (c) $\lambda/2$ and (d) λ

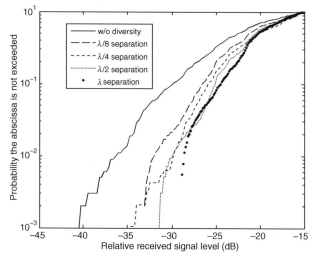

Figure 5.41 CDFs of magnitude of complex envelope without diversity and with selection combining for different antenna separations

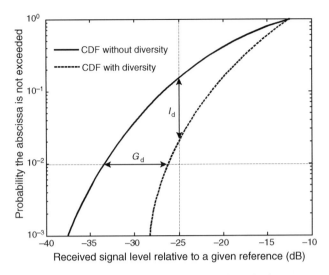

Figure 5.42 Definition of diversity gain and diversity improvement

MS is surrounded by scatterers and is at a distance d from BS. The scatterers about MS can be considered to be within a circle of radius a. The baseline connecting the two antennas makes an angle, ε, with the BS–MS radio path.

As in the previous project, we are interested in finding the spacing, Δd, needed between BS diversity antennas for achieving a sufficiently small cross-correlation coefficient, so that

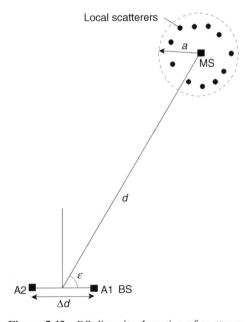

Figure 5.43 BS diversity. Location of scatterers

some degree of diversity gain is feasible. In this case, variables $d, a, \Delta d$ and ε are the relevant parameters as is apparent from the theoretical model provided in [2], i.e.,

$$\rho_{|r|}(\Delta t) \approx J_0^2(z_1) J_0^2(z_2) \tag{5.27}$$

with

$$z_1 = 2\pi f_m \Delta t \, k \sin(\varepsilon) \quad \text{and} \quad z_2 = k^2 \pi f_m \Delta t \sqrt{1 - \frac{3}{4}\cos^2(\varepsilon)} \tag{5.28}$$

where $k = a/d$ is the ratio between the radius of the local scatterers around MS, a, and the BS–MS distance, d. Putting the above expressions in terms of the antenna separation and the mobile speed, we have

$$z_1 = \frac{2\pi}{\lambda} \Delta d \, k \sin(\varepsilon) \quad \text{and} \quad z_2 = k^2 \frac{\pi}{\lambda} \Delta d \sqrt{1 - \frac{3}{4}\cos^2(\varepsilon)} \tag{5.29}$$

In order not to change the structure of the simulators, which are based on `project511`, although these projects are about the uplink, the simulations have been performed for the downlink. Thus, the magnitudes received at MS from BS1 and BS2 have been calculated. These series have been considered to represent their corresponding uplinks.

Figure 5.44(a) illustrates the theoretical and simulated cross-correlation coefficients as a function of their separation (Δd) in wavelengths for three baseline angles, $\varepsilon = 90, 45$ and $15°$. It is clearly shown that the needed spacing between antennas is now much larger, more than an order of magnitude, than in the MS case. It is also apparent how, if MS is in the broadside ($90°$) of the BS array, the needed separation is smaller than for other angles.

Figure 5.44(b) illustrates the CDFs for space diversity with selection combining for different antenna separations. The figure corresponds to the $\varepsilon = 90°$ case. Each CDF represents one point on the $\varepsilon = 90°$ curve in Figure 5.44(a). The diversity gain increases as the separation grows.

The reader is encouraged to modify parameters x, d, a and ε and run new simulations to verify their influence on the diversity gain/improvement. Similarly, the reader is

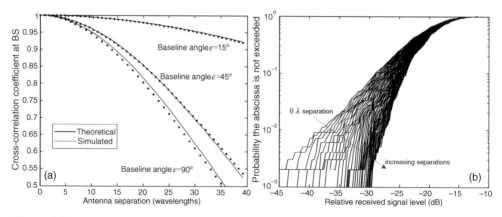

Figure 5.44 (a) Cross-correlation coefficient at BS for different antenna spacings and look-angles $\varepsilon = 90°, 45°$ and $15°$. The radius of the scatterers about MS was 100 m and the BS–MS distance was 20 km. (b) CDFs of selection combined signals for the antenna separation range in (a) and $\varepsilon = 90°$

encouraged to implement other combining methods, especially the maximum ratio technique following [3].

5.3 Summary

In this chapter we have continued our discussion on the narrowband multipath channel. In this case, we have gone one step further to use normalized levels so that we are able to introduce our time series in actual simulators where the working point (the average signal-to-noise ratio) is extremely important. We have continued our discussion using the multiple point-scatterer model for simulating second-order statistics and other parameters. We have then introduced in the model a direct ray, thus generating a Rice series. Afterward, we have presented alternative ways of generating Rayleigh and Rice series: one consisting on an array of low frequency sinusoidal generators; and the other consisting on the combination in quadrature of two random Gaussian, noise-like signals, which are also filtered to force the wanted Doppler characteristics. Finally, we have looked into the issue of space diversity, both at the mobile and the base station sides. The concepts of diversity gain and improvement have been introduced.

References

[1] R.H. Clarke. A statistical theory of mobile radio reception, *BSTJ*, **47**, 1968, 957–1000.

[2] W.C. Jakes. *Microwave Mobile Communication*. John Wiley & Sons, Ltd, Chichester, UK, 1974.

[3] J.D. Parsons (Ed.). *The Mobile Radio Propagation Channel*. John Wiley & Sons, Ltd, Chichester, UK, 2000. Second edition.

[4] C. Loo. Further results on the statistics of propagation data at L-band (1542 MHz) for mobile satellite communications. *41st IEEE Vehicular Technology Conference*, May 19–22, 1991, 51–56.

[5] W.C.Y. Lee. *Mobile Communications Design Fundamentals*. Wiley Series in Telecommunications and Signal Processing. John Wiley & Sons, Ltd, Chichester, UK, 1993.

Software Supplied

In this section, we provide a list of functions and scripts, developed in MATLAB®, implementing the various projects mentioned in this chapter. They are the following:

```
project511              afduration
project512              durationoffades
project513              fCDF
project514              filtersignal
project52               lcr
project53               lcrate
project54               GaussianCDF
project55               rayleigh
project561              RayleighCDF
project562              rayHIST
project563              rice
                        riceCDF
                        ricetheoretical
                        spectrumsignal
```

6

Shadowing and Multipath

6.1 Introduction

In this chapter, the overall narrowband propagation channel is simulated, i.e., we model shadowing plus multipath effects. Additionally, there is the distance-dependent path-loss component (Chapters 1, 2 and 3), which is not accounted for here. The difficulty, in this case, involves taking into consideration the slow and fast signal variations simultaneously. Slow/fast means that the correlation times/distances of the two phenomena are different.

One important issue to be pointed out regarding this chapter is that the reader will become familiar with *mixed distributions*, i.e., combinations of two distributions: a common tool in the field of radiocommunications.

6.2 Projects

The simulations to be performed in this chapter include not only the reproduction of the time-varying amplitudes with their corresponding distributions, but also their phases (complex envelopes) and Doppler spectra.

Project 6.1: The Suzuki Model

This is one of the most common models in classical macrocell mobile system engineering. The underlying assumption is that the direct signal is completely blocked. The received signal is, thus, due to diffuse multipath alone. The average power of the multipath is stationary only over short sections of the MS route while, over longer sections, it varies at a slower rate due to shadowing: short- and long-term statistics.

We have already discussed this issue in Chapter 1 where the concepts of *small area* and *larger area* were presented. Also in Chapter 1, we revised some issues relative to the processing of time series, including the separation of the short- and long-term signal variations. Here, we will try and jointly model the slow and fast variations present in the received signal.

Modeling the Wireless Propagation Channel   F. Pérez Fontán and P. Mariño Espiñeira
© 2008 John Wiley & Sons, Ltd

The above assumptions translate into a combined Rayleigh plus lognormal distribution. This *mixed distribution* is usually called the *Suzuki distribution* [1]. In this case, locally, the signal amplitude variations are Rayleigh distributed with parameter σ, i.e.,

$$f(r|\sigma) = \frac{r}{\sigma^2}\exp\left(-\frac{r^2}{2\sigma^2}\right)(r \geq 0) \tag{6.1}$$

This is valid over short stretches of the MS route, i.e., the so-called *small area*, corresponding to some tens to hundreds of wavelengths. The expression $f(r|\sigma)$ refers to the conditional distribution of r (actually $|r|$) with respect to σ. Parameter σ itself follows a lognormal distribution over larger areas/longer route sections.

The *lognormal distribution* of a random variable, x, is closely linked to that of its logarithm, i.e., $X = \ln(x)$, which follows a normal or Gaussian distribution. Lognormal distributions originate from the multiplication of multiple random variables, similar to what is predicted by the central limit theorem for additive effects, which give rise to a Gaussian distribution. Thus, if the received signal is subjected to a multiplicative concatenation of random attenuators, the resulting signal will be lognormally distributed.

Given a random variable, X, that follows a Gaussian distribution whose pdf is given by

$$p(X) = \frac{1}{\sigma\sqrt{2\pi}}\exp\left[-\frac{1}{2}\left(\frac{X-m}{\sigma}\right)^2\right] \tag{6.2}$$

with mean m and standard deviation σ, if the random variable x is such that $X = \ln(x)$, then x follows a lognormal distribution with pdf

$$p(x) = \frac{1}{\sigma\sqrt{2\pi}}\frac{1}{x}\exp\left[-\frac{1}{2}\left(\frac{\ln x - m}{\sigma}\right)^2\right] \tag{6.3}$$

As shown in the above formula, the parameters of the associated Gaussian distribution are often quoted as the parameters for the lognormal distribution. The actual parameters of the lognormal distribution are somewhat more complicated, Table 6.1.

This distribution will allow us to describe powers, voltages or field strengths in linear units that, when represented in logarithmic units, are normally distributed.

We are not interested in voltages or powers expressed in Nepers but in dB. This means that we have to slightly change the expression of the lognormal pdf to account for the use of decimal logarithms, and a 10 or 20 multiplicative factor, depending on whether the

Table 6.1 Parameters of the lognormal distribution

Most probable value (mode)	$\exp(m - \sigma^2)$
Median value	$\exp(m)$
Mean value	$\exp(m + \sigma^2/2)\sqrt{\exp(\sigma^2) - 1}$
Root mean square value	$\exp(m + \sigma^2)$
Standard deviation	$\exp(m + \sigma^2/2)$

magnitude of interest is a power or a voltage. Thus, the lognormal distribution, when the variable is the *power*, is given by

$$f(p) = \frac{4.343}{p\sqrt{2\pi}\Sigma} \exp\left\{-\frac{[10\log p - M]^2}{2\Sigma^2}\right\} \tag{6.4}$$

while, when it is the *voltage*, the pdf is

$$f(v) = \frac{8.686}{v\sqrt{2\pi}\Sigma} \exp\left\{-\frac{[20\log v - M]^2}{2\Sigma^2}\right\} \tag{6.5}$$

where, now, the mean, M, and standard deviation, Σ, are in dB.

Coming back to the Suzuki distribution, parameter σ, assumed constant over a small area, varies according to a lognormal distribution over larger areas, i.e.,

$$f(\sigma) = \frac{8.686}{\Sigma\sigma\sqrt{2\pi}} \exp\left[-\frac{(20\log\sigma - M)^2}{2\Sigma^2}\right] \tag{6.6}$$

Writing the whole expression, and eliminating variable σ through integration, gives

$$\begin{aligned}
f(r) &= \int_0^\infty f(r|\sigma)f(\sigma)d\sigma \\
&= \frac{8.686r}{\Sigma\sqrt{2\pi}} \int_0^\infty \frac{1}{\sigma^3} \exp\left[-\frac{(20\log\sigma - M)^2}{2\Sigma^2}\right] \exp\left(-\frac{r^2}{2\sigma^2}\right) d\sigma \ (r \geq 0)
\end{aligned} \tag{6.7}$$

This is a rather involved expression that must be integrated numerically using MATLAB® (MATLAB® is a registered trademark of The MathWorks, Inc.) function `quad` (`project612`). The Suzuki distribution, which has two parameters, reduces to a Rayleigh distribution when the standard deviation is zero, i.e., $\Sigma = 0$. We have already come across parameters M and Σ, they are related to the *larger area mean* and *location variability* discussed in Chapters 1 and 3.

We are interested in generating Suzuki distributed time series (`project611`). For this, we can take into account its phasor representation, i.e.,

$$r_T \exp(j\phi_T) = r_M \exp(j\phi_M) \tag{6.8}$$

with T meaning total and M multipath, and where the main difference with respect to the Rayleigh generator (Project 5.3) is that the parameter σ is no longer constant, but varies according to a lognormal distribution, and its coherence time/distance is larger than that of the Rayleigh variations, i.e., it varies more slowly.

The schematic diagram of the series generator is shown in Figure 6.1 where a Clarke/Jakes [2][3] spectrum is depicted (Chapter 5). The main difference resides in the fact that σ is driven by a lognormal series generator. In turn, this is based on a Gaussian generator plus an antilogarithm operation for converting the series to lognormal.

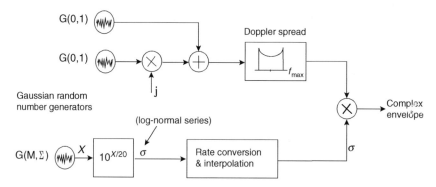

Figure 6.1 Schematic diagram of Suzuki time series generator

Figure 6.1 shows the circuit representation for the Suzuki model: two zero-mean and unit-standard deviation Gaussian series in quadrature go through a unit-gain, low-pass, U-shaped filter thus simulating the Clarke/Jakes spectrum.

After Doppler shaping, the resulting complex series is multiplied by σ, with $2\sigma^2$ being the rms squared value of the associated Rayleigh distribution that describes the variations due to diffuse multipath. In this case, the value of σ varies slowly as MS moves. How the variations of σ were generated is discussed next.

The sampling frequency selection must be based on that required for the fastest of the two processes, i.e., the multipath fading. As discussed in Project 5.3, at least twice the maximum Doppler frequency, f_m, is required. We specify the sampling rate in terms of a fraction, F, of the wavelength.

This minimum fraction of the wavelength needed can be explained by expressing the sample spacing, d_s, in the time domain taking into account the MS speed, V. Thus, $t_s = d_s/V$. We also know that $d_s = \lambda/F$ then, the sampling frequency (Hz) is given by

$$f_s = \frac{V}{d_s} = \frac{FV}{\lambda} = F f_{max} \qquad (6.9)$$

and we know from the sampling theorem that F must be greater than 2.

For generating the slower variations, we used the procedure discussed in Chapter 3, where uncorrelated Gaussian samples are spaced by one correlation length, and then correlated samples are interpolated in between (see `project312`).

The series from the two rails (slow and fast variations), sampled at the same rate, can now be multiplied to produce a Suzuki distributed time series, that preserves both the rates of change of the shadowing and multipath phenomena, and the Doppler spectra of the multipath. The slower variations will show a much narrower spectrum, for which no well-recognized model is available. In any case, this narrower spectral component will be unnoticeable within the wider multipath spectrum.

In `project611` we chose the following settings: the carrier was set to 2 GHz, the mobile speed to 10 m/s, the sampling rate was defined by the number of samples per wavelength, F, equal to four in this case. The mean and standard deviation of the Gaussian associated to the lognormal variations were set to -13 dB and 3.4 dB (location variability), the correlation length was set to 9 m.

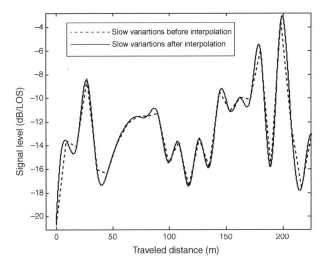

Figure 6.2   Slow variations in dB: before and after interpolation

Figure 6.2 illustrates the slow variations before and after interpolation, i.e., samples were first generated (`randn`) with a spacing of L_{corr} m. These samples are uncorrelated. Then, they were interpolated to a rate λ/F. Figure 6.3 shows the implemented FIR version of Clarke's Doppler filter. A very simple approach has been used for creating the Doppler filter, where the wanted U-shaped frequency response has been sampled and then, the inverse FFT taken. This creates an FIR version of the Doppler filter with a very large number of coefficients. Much more efficient implementations are possible in the form of IIR filters with a limited number of coefficients. The creation of such implementation is beyond the scope of this book and is left for the reader to look into.

Figure 6.4(a) illustrates the fast variations for $\sigma = 1$ and Figure 6.4(b) shows the Doppler spectrum of the filtered Rayleigh variations. Figure 6.5(a) illustrates the overall slow plus fast variations together with the interpolated slow variations. Figure 6.5(b) shows the overall

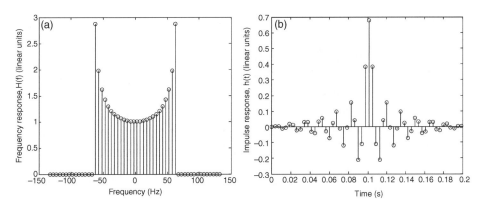

Figure 6.3   (a) Approximation of Jakes/Clarke Doppler filter in the frequency domain. (b) Approximation in the time domain

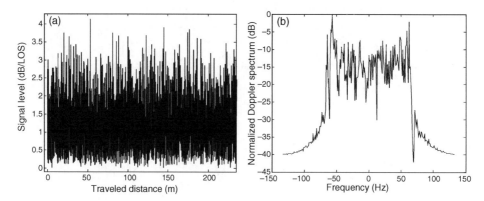

Figure 6.4   Fast variations. (a) Magnitude. (b) Doppler spectrum

absolute phase, where no clear trend can be observed, consistent with the Rayleigh model where no direct signal is present. Figure 6.6(a) illustrates the CDF of the overall synthesized series according to the Suzuki model together with the CDF of the lognormal series.

After running `project611`, we can call up `project612` to calculate the theoretical Suzuki CDF and plot it together with the series CDF (Figure 6.6(b)). The plots are in dB, and the differences observed are due to the fact that the two components, fast and slow, vary at fairly different rates. A much longer series is needed to achieve stable statistics.

Project 6.2: Power Control

This project uses a modified version of `project611` to simulate a very simple *power control* (PC) scheme. Power control is extremely important in all radio systems, since it reduces the transmitted power to the minimum necessary for a communication link to perform correctly, while keeping interference levels to a minimum. In *direct sequence spread spectrum* systems (DS-SS) such as IS-95 or 3G systems, the accuracy of PC mechanisms is vital for a correct

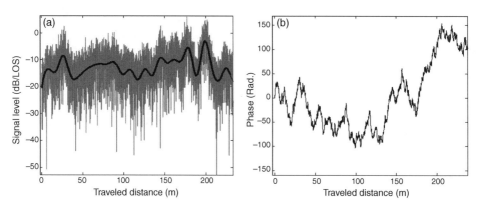

Figure 6.5   (a) Overall Suzuki variations and slow lognormal variations. (b) Phase variations of overall signal

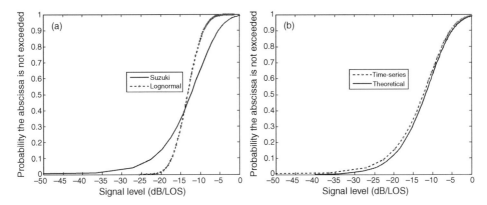

Figure 6.6 (a) CDFs of Suzuki and lognormal series. (b) CDFs of simulated series and corresponding theoretical Suzuki distribution

operation of such systems, and high power control update rates are needed. For example, in 3G systems a rate of 1500 updates per second is used. This allows countering not only the shadowing effects but also the fast fade effects, especially if the mobile speed is not too high.

The implementations in `project621` (for a mobile speed of 10 m/s, i.e., 36 km/h) and `project622` (for a mobile speed of 56 m/s, approximately 200 km/h) assume an instantaneous response PC system, consisting of a *return link* through which commands are sent to the other end to increase or decrease its transmit power by 1 dB, 1500 times per second. This is not a realistic scenario, since there is a propagation or flight time, and a processing time at both ends, which here have been assumed to be negligible.

Moreover, the processing normally includes prediction filters which try to extrapolate the behavior of the received signal for optimizing the PC process. We have assumed that, at the receive side, a sliding window of duration 1/1500 s, the inverse of the update rate, is used for deciding which command to send back to the transmitter. The result of the averaging process is compared with a preset *threshold*. A result above the threshold means sending back a 'decrease power' order, and vice versa.

Another important change with respect to `project611` is that now we have created a time axis from the traveled distance axis. This is needed because the averaging operations performed at the receiver are obviously carried out in time, and encompass route sections of different lengths depending on the MS speed. Furthermore, to achieve a sufficiently fine characterization of the series in the time domain, we have used a large number of samples per wavelength, *F*, in the simulations.

The settings in `project621` and `project622` are as follows: correlation distance, `Lcorr=9.0`, sampling fraction of wavelength, `F=500`, large area average, `M=−13`, location variability, `S=6.0`. Only the MS speeds are different as mentioned above. The threshold was set to the median of the original received signal to help in the CDF comparisons.

Figure 6.7 shows the received signal without power control. This series was generated using the Suzuki model. Also in the figure, the slow variations are plotted. Figure 6.8(a) shows the same series together with the one received under PC conditions. The figure also shows the 1 dB increments or decrements in the power-controlled transmit power, following the orders sent back from the receiver to the transmitter. Figure 6.8(b) shows a zoomed-in section of the previous figure where we can appreciate the discrete nature of both the

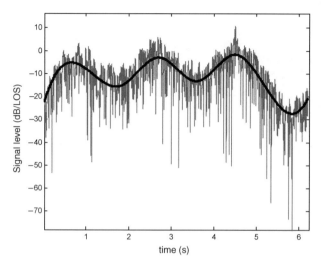

Figure 6.7 Slow speed case. Original series, Suzuki distributed and slow variations

transmitted power, and the power-controlled received signal. Finally, we show in Figure 6.9 the CDFs of the power-controlled received signal and the original uncontrolled signal. Observe the much smaller dynamic range in the controlled signal. These figures were obtained with `project621`, which corresponds to a moderate vehicular speed of 10 m/s.

We followed the same approach in `project622`, where we assumed a much larger speed of 56 m/s corresponding, for example, to a receiver on a high speed train. We repeat the same types of plots – Figure 6.10 shows another received signal (remember that we use three random number generators). However, we have simulated the same route length which now corresponds to a much shorter time.

Figure 6.11(a) shows the original signal, the controlled transmitted power and the corresponding controlled received power. We can see that, as the speed is much faster,

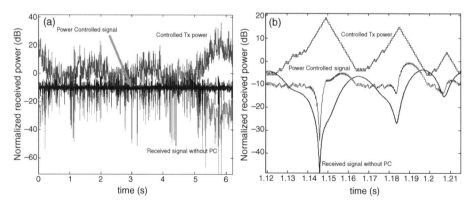

Figure 6.8 (a) Slow speed case. Original uncontrolled series, controlled transmitted power and controlled received power. (b) Zoomed-in plot

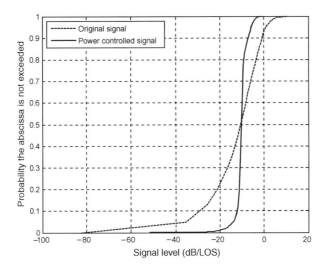

Figure 6.9 Slow speed case. CDFs of original and power-controlled received signals

the PC mechanism is not able to follow and correct for the fast variations while the slower variations are fairly well compensated. Figure 6.11(b) shows a closer look at a section of the previous figure. Finally, Figure 6.12 shows the corresponding CDFs, where the deficient compensation is clearly observed. Notice how the statistics of the two signals are fairly close, indicating the lack of correction.

The reader is encouraged to try other thresholds and develop more complex algorithms to gain a better insight into power control issues.

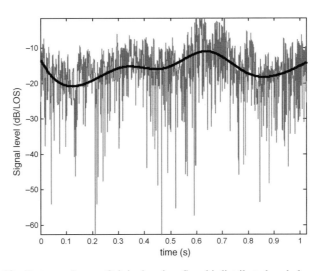

Figure 6.10 Fast speed case. Original series, Suzuki distributed and slow variations

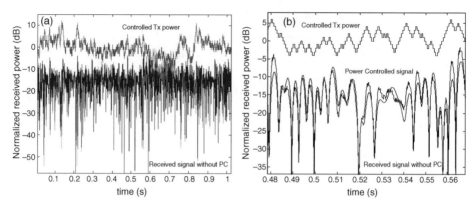

Figure 6.11 (a) Fast speed case. Original uncontrolled series, controlled transmitted power and controlled received power. (b) Zoomed-in plot

Project 6.3: The Loo Model

The Suzuki distribution represents a worst-case scenario, which is frequently used in macrocells as a benchmark for system performance, and BS deployment studies. It assumes that the direct signal is totally blocked. Here, we are going to try and simulate a much milder channel, in the sense that the direct signal is supposed to be present, being either totally unblocked or partially shadowed. The Loo model is another mixed distribution that, given that it has three parameters, instead of two as Suzuki's, is more versatile, at the expense of requiring a more complex procedure for parameter extraction from measured data.

This distribution has been frequently used in the modeling of the land mobile satellite channel (Chapter 9). The conditions of this model are typically found when there exists a slowly varying (due to shadowing) direct signal together with a constant-power diffuse multipath component. This can be modeled using a combination of a Rice and a lognormal

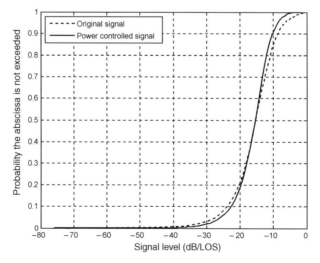

Figure 6.12 High speed case. CDFs of original and power-controlled received signals

distribution. Here, again, we will be synthesizing complex time series (project631) where the multipath part varies faster than the shadowing part.

The *Loo distribution* [4] is based on the Rice distribution. The received signal variations (voltage or field strength) are assumed to vary according to a Ricean distribution with parameters a and σ, i.e.,

$$f(r|a) = \frac{r}{\sigma^2}\exp\left(-\frac{r^2 + a^2}{2\sigma^2}\right)I_0\left(\frac{ra}{\sigma^2}\right)(r \geq 0) \tag{6.10}$$

This is valid for short sections of the traveled route. For longer sections, the direct signal's amplitude is assumed to vary according to a lognormal distribution, i.e.,

$$f(a) = \frac{8.686}{\Sigma a\sqrt{2\pi}}\exp\left[-\frac{(20\log a - \mathrm{M})^2}{2\Sigma^2}\right](r \geq 0) \tag{6.11}$$

where M and Σ are the mean and standard deviation of the normal distribution (associated to the lognormal) for the direct signal's amplitude $(A = 20\log(a))$ expressed in dB.

Thus, the Loo distribution assumes that the received complex envelope consists of the coherent sum of two phasors,

$$r_\mathrm{T}\exp(j\phi_\mathrm{T}) = r_\mathrm{D}\exp(j\phi_\mathrm{D}) + r_\mathrm{M}\exp(j\phi_\mathrm{M}) \tag{6.12}$$

where subscript T means total or overall, D direct signal and M indicates diffuse multipath. As said above, the direct signal's amplitude is lognormally distributed due to shadowing and the multipath component's parameter, σ, remains constant throughout. The overall distribution, i.e., for longer stretches of the route is given by

$$
\begin{aligned}
f(r) &= \int_0^\infty f(r|a)f(a)\,da \\
&= \frac{8.686r}{\sigma^2\Sigma\sqrt{2\pi}}\int_0^\infty \frac{1}{a}\exp\left(-\frac{r^2 + a^2}{2\sigma^2}\right)\exp\left[-\frac{(20\log a - \mathrm{M})^2}{2\Sigma^2}\right]I_0\left(\frac{ra}{\sigma^2}\right)da\ (r \geq 0)
\end{aligned}
\tag{6.13}
$$

with M and Σ in dB and where $MP(\mathrm{dB}) = 10\log 2\sigma^2$ is the rms squared value of the multipath component expressed in dB.

The Loo distribution is very versatile as it includes, as special cases, the normal and the Rice distributions for large values of a, and the Rayleigh distribution for negligible values of a. This property makes this distribution valid for a very wide range of conditions spanning from line-of-sight (LOS) to heavily shadowed conditions.

To simulate Loo-distributed time series, the schematic diagram in Figure 6.13 was followed when implementing project631. Additionally, phase and Doppler spectra characteristics are also introduced in the simulated complex envelope, $r_\mathrm{T}(t)\exp[j\phi_\mathrm{T}(t)]$.

Figure 6.13 shows the circuit representation for the Loo series generator: two zero-mean and unit-standard deviation Gaussian series in quadrature are passed through a unit-energy,

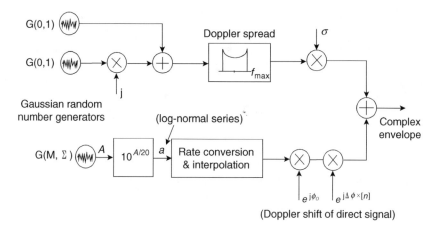

Figure 6.13 Circuit implementation of Loo model with Clarke/Jakes Doppler shaping

low-pass, U-shaped filter simulating Clarke's spectrum. After Doppler shaping, the resulting complex series is multiplied by σ.

The lower rail in the schematic diagram in Figure 6.13 performs the simulation of the direct signal's amplitude and phase variations. In a first step, an M (dB) mean and Σ (dB) standard deviation Gaussian series is generated. The series is converted to linear units (lognormal distributed) by computing the antilogarithm, $a = 10^{A/20}$.

The direct signal's amplitude is subjected to variations due to shadowing that are slower than those due to multipath. The rate of change of these slow variations is characterized by the *correlation time*, t_{corr}, parameter or, in the traveled distance domain, the *correlation length*, L_{corr}. Typically, correlation lengths for these slow variations depend on the size of the objects causing the blockages, i.e., several meters. The lognormal series generator (lower rail in Figure 6.13) produces uncorrelated samples that are spaced L_{corr} meters.

At the same time, the fast variations were generated with a much higher sampling rate to account for the Doppler bandwidth, i.e., $\pm f_{max} = \pm V/\lambda$. This means that a combined rate conversion plus interpolation process must be applied to the slower variations, thus allowing the complex addition of the direct and the multipath phasors, performed at the end of the simulation chain. Here, the sampling is performed in the traveled distance using a spacing of a fraction, F, of the wavelength.

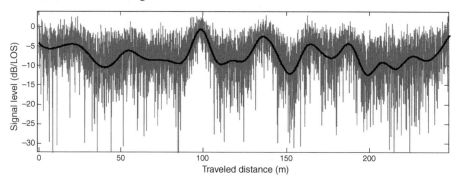

Figure 6.14 Overall Loo-distributed variations and direct signal variations

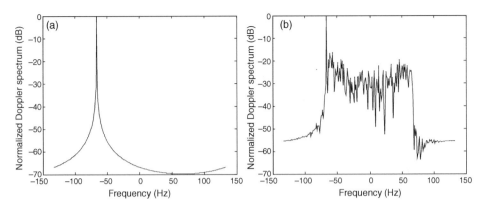

Figure 6.15   (a) Doppler spectrum of direct signal. (b) Doppler spectrum of Loo-distributed complex series

In the next step, the phase variations in the direct signal are introduced. These are assumed to be linearly varying giving rise to a constant Doppler spectral line, which depends on the MS velocity and the angle of arrival, azimuth and elevation, with respect to the MS trajectory, i.e.,

$$f_{\text{Direct signal}} = (V/\lambda) \cos \varphi \cos \theta = f_{\max} \cos \varphi \cos \theta \qquad (6.14)$$

with φ and θ being the relative azimuth, with respect to the MS route, and elevation angles, respectively. We restricted ourselves to the horizontal plane, and dropped the dependence on angle θ, i.e.,

$$f_{\text{Direct signal}} = (V/\lambda) \cos \varphi = f_{\max} \cos \varphi \qquad (6.15)$$

The constant phase increment in Figure 6.13 is given by (Project 4.1)

$$\Delta\phi = 2\pi \frac{\cos(\varphi)}{F}. \qquad (6.16)$$

Parameter $\Delta\phi$ is multiplied by the sample number, $[n]$, to generate the current phase increment. In the schematic diagram in Figure 6.13 an optional initial phase term, ϕ_0, is also shown.

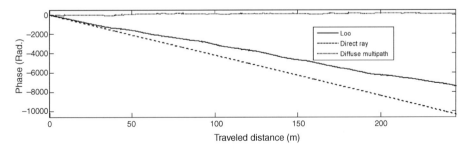

Figure 6.16   Absolute phases of Loo, direct and multipath contributions

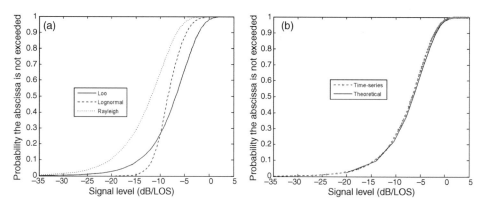

Figure 6.17　(a) Overall Loo series, lognormal and Rayleigh CDFs. (b) Loo series and theoretical CDFs with levels in dB

The settings used for running `project631` are given next. The Loo model parameters were `MP=-10`, `M=-8.0` and `S=3.0`, the correlation length was `Lcorr=9.0`. The angle of arrival of the direct signal was set to `AoA=180`. Figure 6.14 shows the synthesized Loo series together with the slow direct signal. Observe the changing character of the resulting series, that is, for the larger values of the direct signal, the overall series behaves more like a Rice series, whereas for the smaller values of the direct signal, the overall series shows Rayleigh characteristics.

Figure 6.15(a) illustrates the spectrum of the direct signal showing that it arrives from behind. Figure 6.15(b) shows the overall signal's Doppler spectrum where the U-shaped Clarke spectrum is present together with the direct signal part. Figure 6.16 illustrates the overall absolute phase and the phases of the two components, where it is clearly shown how the direct signal component dominates over the multipath contributions. The figure shows how small phase variations due to multipath are superposed on the dominating direct signal phase. Finally, Figure 6.17(a) depicts the CDFs of the overall signal and of its components.

If `project632` is run immediately after `project631`, i.e., keeping the contents of the MATLAB® workspace, Figure 6.17(b) is produced where the theoretical Loo CDF, and that corresponding to the synthetic series are plotted together in logarithmic units. The small differences, as in the Suzuki case, are due to the length of the generated series not being long enough.

The reader is encouraged to modify the scripts provided and try different parameters to get a feel for their relative influence on the resulting series.

6.3 Summary

In this chapter we conclude our analysis of the narrowband channel by modeling together the slow and fast variations due to shadowing and multipath, respectively. We have also presented two mixed distributions: Suzuki and Loo. We have also simulated, in a very simplistic way, the power control mechanism. In the next chapter, we go one step further and introduce the wideband characteristics of the multipath propagation channel.

References

[1] H. Suzuki. A statistical model for urban radio propagation. *IEEE Trans. Comm.*, **25**(7), 1979, 213–225.

[2] R.H. Clarke. A statistical theory of mobile radio reception. *BSTJ*, **47**, 1968, 957–1000.

[3] W.C. Jakes. *Microwave Mobile Communication.* John Wiley & Sons, Ltd, Chichester, UK, 1974.

[4] C. Loo. A statistical model for land mobile satellite link. *IEEE Trans. Vehic. Tech.*, **34**, 1985, 122–127.

Software Supplied

In this section, we provide a list of functions and scripts, developed in MATLAB®, implementing the various projects mentioned in this chapter. They are the following:

```
project611             fCDF
project612             jackes
project621             jackeswofigs
project622             lootheoretical
project631             ricetheoretical
project632             suzukitheoretical
```

7

Multipath: Wideband Channel

7.1 Introduction

In addition to fading, the mobile channel gives rise to time dispersion. Angle dispersion also occurs but it will be dealt with in Chapter 10. Time dispersion translates, in the frequency domain, to non-uniform filtering of the transmitted signal, i.e., a non-ideal response (Chapter 1), meaning a non-flat magnitude and nonlinear phase. Moreover, this response will vary in time: a *frequency-selective fading channel*. The wideband condition is a relative one that depends on the transmitted signal bandwidth. Anti-multipath techniques such as equalization, multi-carrier orthogonal frequency-division multiplexing (OFDM) or code division multiple access, direct sequence-spread spectrum (CDMA DS-SS) may be used to counter these effects.

In this chapter, several aspects of the wideband propagation channel are studied. The *wide sense stationary-uncorrelated scattering* (WWSUS) assumption is discussed. How the channel response can be parameterized in terms of its amplitude and phase responses is illustrated by means of channel transfer functions, created using the multiple point-scatterer model. Some of the channel functions identified by Bello [1] are presented and visualized using simulation results. Issues relative to channel simulation, including the time-varying *tapped delay-line* (TDL) model, are presented. How to create TDL time series is discussed as well.

In previous chapters, we have assumed that an unmodulated carrier (CW) was transmitted. This is practically equivalent to studying the channel effects on signals with narrow bandwidths. However, it is very important to quantify the effect of the channel on any kind of transmitted signal, i.e., signals that occupy a wide range of possible bandwidths.

Assuming two frequencies belonging to a transmitted signal with a given bandwidth, if these two frequencies are close, the different propagation paths (multipath) will have approximately the same electrical lengths: $2\pi d/\lambda_1$ and $2\pi d/\lambda_2$. This means that their corresponding received amplitudes and phases will vary in time, in approximately the same way. In this case, we would have *flat fading* conditions (Figure 7.1(a)).

As the frequency separation increases, the fading behavior at one of the frequencies tends to be uncorrelated with respect to the other. This is due to the fact that the electric lengths

Modeling the Wireless Propagation Channel   F. Pérez Fontán and P. Mariño Espiñeira
© 2008 John Wiley & Sons, Ltd

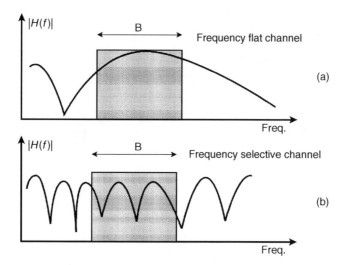

Figure 7.1   Channel frequency transfer function vs. transmitted signal bandwidth

will be significantly different. The correlation between the behavior at these two frequencies will depend on the *time spreading* caused by the environment (dispersion of path lengths and associated powers). Thus, signals occupying larger bandwidths will be distorted (Figure 7.1(b)). This phenomenon is known as *frequency selective fading*, which means that there is a non-uniform frequency response throughout the transmitted bandwidth. The minimum bandwidth for which selective fading effects start to be observed is known as the channel's *coherence bandwidth*.

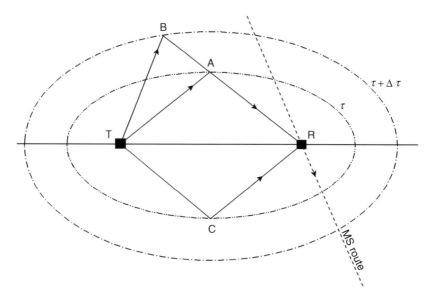

Figure 7.2   Ellipses defining scattering points with equal delays [2]

The delayed replicas of the transmitted signal reaching MS may be associated with specific scatterers in the traversed environment [2]. To fully characterize the propagation channel at a given point on the MS route, knowing the powers and delays of the various echoes is not sufficient, it is also necessary to measure their angles of arrival and departure. One way of assessing the angles of arrival is by keeping track of the received Doppler spectrum. As we have seen in Chapter 4, there is a direct relationship between Doppler shift and angle of arrival.

Assuming a simple model where only single scattered contributions are taken into account, the echoes with a given delay will be located on an ellipse whose focal points are the transmitter, T, and the receiver, R, positions (Figure 7.2) [2]. By considering three scatterers A, B, C, paths TAR and TBR may be resolved (in spite of having the same Doppler) by their different delays. Furthermore, it is possible to distinguish between TAR and TCR, which have the same delay, by their Dopplers. The angles of arrival, α_i, may be obtained by measuring the Doppler shift, $f_m \cos(\alpha_i)$, except for a left–right ambiguity.

7.2 Deterministic Multiple Point-Scatterer Model

It is interesting to try and model the behavior of the channel in a deterministic way [2][3]. The magnitudes of the various scattered contributions may be expressed in terms of their radar cross-sections, $\sigma_i(\mathrm{m}^2)$. Similarly, other propagation mechanisms like specular reflections may be introduced, which are characterized by their reflection coefficients, as discussed in Chapter 1. Through this very simple approach, a physical insight into the main features of the channel can be obtained. Assuming the direct signal is totally blocked, and only scattered contributions reach the receiver, the received field strength, e_T, is given by

$$e_T \; \alpha \; \sum \frac{\sqrt{\sigma_i}}{d_{i-1} \cdot d_{i-2}} \exp\left[-j \frac{2\pi}{\lambda} (d_{i-1} + d_{i-2}) \right] \exp(j\xi_i) \qquad (7.1)$$

where α means proportional to. The *radar cross-section* is a power-related concept where the phase information is not available. Here, however, we include a phase term, $\exp(j\xi_i)$, for completeness. We have frequently neglected this term in previous and ensuing simulations, and only used it when strictly necessary, given that the movement of MS already causes phase variations due to changes in the sub-path distances, d_{i-1} and d_{i-2}.

Assuming, for example [4], that the propagation environment described schematically in Figure 7.3 is traversed by MS, the contributions shown in the figure are received. These echoes are classified as a function of their delays and Dopplers. We will learn later on that this representation is normally called the *channel scattering function*. Note that the direct signal produces a strong peak with a relative delay equal to zero and negative Doppler equal to $-f_m$. A strong, far echo from the hills is also observed in the direction of the MS route, giving rise to a positive Doppler equal to f_m. There is also a zero Doppler echo from the group of trees perpendicular to the route and, finally, one last echo comes from another group of trees with negative Doppler.

In previous chapters we have assumed that, even though the differences in radio path lengths for the various scatterers gave rise to phase changes, their associated delays were not significant. Now, we include the delay in the modeling, as it is the source of time spreading,

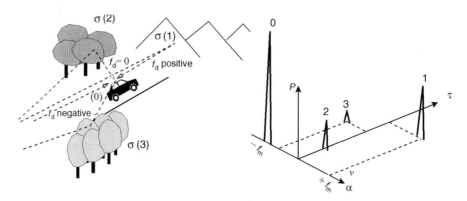

Figure 7.3   Example mobile scenario and corresponding scattering function [4]

an essential component of the mobile propagation channel, together with Doppler and fading.

Let us assume a *time-invariant channel* for a fixed MS position [5]. The relationship between the transmitted signal, in its pass-band (RF) representation, and its associated *complex envelope* (CE) representation, is given by

$$x_{\mathrm{RF}}(t) = \mathrm{Re}[x_{\mathrm{CE}}(t)\exp(\mathrm{j}2\pi f_0 t)] \tag{7.2}$$

If this signal is transmitted through a static scattering medium, the received signal is given by

$$y_{\mathrm{RF}}(t) = \mathrm{Re}\left\{\sum_{i=1}^{N}\tilde{a}_i x_{\mathrm{CE}}(t-\tau_i)\exp[\mathrm{j}2\pi f_0(t-\tau_i)]\right\} = \mathrm{Re}[y_{\mathrm{CE}}(t)\exp(\mathrm{j}2\pi f_0 t)] \tag{7.3}$$

where

$$y_{\mathrm{CE}}(t) = \sum_{i=1}^{N}\tilde{a}_i \exp(-\mathrm{j}2\pi f_0\tau_i)x_{\mathrm{CE}}(t-\tau_i) \tag{7.4}$$

Complex coefficient $\tilde{a}_i$ represents the magnitude and phase of the contribution from scatterer i, and τ_i is its associated propagation time (delay). Delays can be interchanged with path lengths using $\tau_i = d_i/c$, thus,

$$y_{\mathrm{CE}}(t) = \sum_{i=1}^{N}\tilde{a}_i \exp\left(-\mathrm{j}\frac{2\pi}{\lambda_0}d_i\right)x_{\mathrm{CE}}(t-d_i/c) \tag{7.5}$$

From the above, the *channel impulse response* in the CE (low-pass equivalent) domain is given by

$$h_{\mathrm{CE}}(\tau) = \sum_{i=1}^{N}\tilde{a}_i \exp(-\mathrm{j}2\pi f_0\tau_i)\delta(\tau-\tau_i) \tag{7.6}$$

Note that there are two phase terms involved, one due to the actual scattering phenomenon, contained in $\tilde{a}_i$, and the second due to the propagation path distance. For example, if we were talking about a reflected ray, the first phase term would correspond to the phase of the reflection coefficient.

When MS moves [5], the individual radio paths, and their associated delays, change in time. A common assumption is that each individual path length changes at a constant rate, depending on the angle of incidence, α_i, with respect to the direction of movement of MS (Project 4.1), i.e., $d_i'(t) = d_i - V\cos(\alpha_i)t = d_i + d_i(t)$ or, alternatively, $\tau_i'(t) = \tau_i - V\cos(\alpha_i)t/c = \tau_i + \tau_i(t)$. In this case, the received signal can be put as

$$y_{CE}(t) = \sum_{i=1}^{N} \tilde{a}_i \exp\{-j2\pi f_0[\tau_i + \tau_i(t)]\}x_{CE}\{t - [\tau_i + \tau_i(t)]\} \tag{7.7}$$

The changing delays in the complex exponential can be attributed to the Doppler effect, i.e.,

$$
\begin{aligned}
\exp[-j2\pi f_0\tau_i(t)] &= \exp\left[j2\pi f_0 \frac{V\cos(\alpha_i)t}{c}\right] \\
&= \exp\left[j2\pi \frac{V}{\lambda_c}\cos(\alpha_i)t\right] = \exp[j2\pi f_{Di}t]
\end{aligned}
\tag{7.8}
$$

Finally, the *time-varying impulse response* is given by

$$h_{CE}(t,\tau) = \sum_{i=1}^{N} \tilde{a}_i \exp\{-j2\pi f_0[\tau_i + \tau_i(t)]\}\delta\{\tau - [\tau_i + \tau_i(t)]\} \tag{7.9}$$

and the *time-varying frequency response* is

$$T_{CE}(t,f) = \sum_{i=1}^{N} \tilde{a}_i \exp\{-j2\pi(f + f_0)[\tau_i + \tau_i(t)]\} \tag{7.10}$$

Note that, following [2], the time-varying frequency response is called T (see next section) instead of H.

It is important to have a clear picture of what a time-varying channel response means [5]. In the case of a time-invariant channel, function $h(\tau)$ is the response observed at an arbitrary time, t, to an impulse transmitted τ seconds earlier. This response does not depend on the observation time, only on the delay, τ, between the times of transmission and observation. If the channel is time varying, $h(t,\tau)$ depends on the observation time, t, and on the delay, τ. The response, $h(t,\tau)$, is now observed at time t and the impulse is applied τ seconds earlier, that is, at time $t - \tau$. If we particularize the response for the time-invariant case, $h(t,\tau)$ will no longer depend on the observation time, t, becoming $h(0,\tau) = h(\tau)$.

7.3 Channel System Functions

We have already seen how the input and output signals can be linked in the frame of the multiple point-scatterer model. Through this approach, we have been able to find two channel system functions, h and T. In total, four system functions can be defined, namely, $h(t, \tau)$, $T(t, f)$, $H(f, \nu)$ and $S(\tau, \nu)$, where ν is the Doppler shift [2]. First, we consider a *time-varying deterministic channel*. Later, the random case will be introduced.

The time-domain description of a linear system is specified by its impulse response. Since the channel is time-variant, the impulse response is also a time-varying function. If the low-pass equivalent, *time-varying impulse response* is $h(t, \tau)$, where τ is the delay variable, then the complex envelope at the receive side, $y(t)$, is related to the input, $x(t)$, by a *convolution integral*, i.e.,

$$y(t) = \int_{-\infty}^{\infty} x(t - \tau)h(t, \tau)d\tau \tag{7.11}$$

where the time variations are contained in the fluctuations of $h(t, \tau)$. The convolution can be rewritten as a summation, i.e.,

$$y(t) = \Delta\tau \sum_{m=1}^{n} x(t - m\Delta\tau)h(t, m\Delta\tau) \tag{7.12}$$

This expression provides a physical representation (Figure 7.4(a)) of the channel in the form of a dense tapped delay-line (TDL) composed of differential delay elements and modulators [2].

The channel can also be represented by another function, $H(f, \nu)$, dual of $h(t, \tau)$, which links the output to the input spectra, i.e.,

$$Y(f) = \int_{-\infty}^{\infty} X(f - \nu)H(f - \nu, \nu)d\nu \tag{7.13}$$

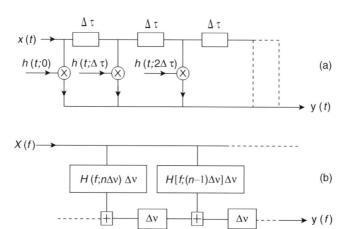

Figure 7.4 (a) Dense TDL channel representation. (b) Channel representation in the frequency domain

Again, the equation can be written as a summation,

$$Y(f) = \Delta v \sum_{m=1}^{n} X(f - m\Delta v)H(f - m\Delta v, m\Delta v) \tag{7.14}$$

This function allows a further physical representation of the channel in the form of a dense frequency conversion chain, analogous to a TDL, consisting of a bank of filters with transfer functions, $H(f, v)\Delta v$, followed by Doppler shifting frequency converters, producing Doppler shifts in the range $(v, v + \Delta v)$ Hz (Figure 7.4(b)).

Another way of representing the channel is by using the *time-varying transfer function*, $T(t, f)$, i.e.,

$$y(t) = \int_{-\infty}^{\infty} X(f)T(f, t)\exp(j2\pi ft)df \tag{7.15}$$

Function $T(t, f)$ is the Fourier transform of $h(t, \tau)$ with respect to variable τ, and also the inverse Fourier transform of $H(f, v)$ with respect to variable v [2], i.e.,

$$T(f, t) = \int_{-\infty}^{\infty} h(t, \tau)\exp(-j2\pi f\tau)d\tau = \int_{-\infty}^{\infty} H(f, v)\exp(j2\pi vt)dv \tag{7.16}$$

Functions h and H describe the channel dispersive behavior in terms of only one of the dispersion variables: τ or v. Another important representation of the channel is by means of the so-called *scattering function*, $S(\tau, v)$, which includes both dispersion variables. Function $h(t, \tau)$ can be obtained as the inverse Fourier transform of $S(\tau, v)$, i.e.,

$$h(t, \tau) = \int_{-\infty}^{\infty} S(\tau, v)\exp(j2\pi vt)dv \tag{7.17}$$

and the input–output relationship using $S(\tau, v)$ is given by

$$y(t) = \int_{-\infty}^{\infty} \int_{-\infty}^{\infty} x(t - \tau)S(\tau, v)\exp(j2\pi vt)dvd\tau \tag{7.18}$$

This equation shows that the output can be represented as the sum of delayed and Doppler-shifted contributions whose differential scattering amplitudes are given by $S(\tau, v)dvd\tau$ [2]. Thus, $S(\tau, v)$ explicitly describes the dispersive behavior of the channel in terms of time delays and Doppler shifts, and can be physically interpreted with reference to the geometry presented in Figure 7.2. In the frame of the multiple point-scatterer model, the scattering function can be put in the form

$$S(\tau, v) = \sum_{i=1}^{N} \tilde{a}_i \delta(\tau - \tau_i)\delta(v - v_i) \tag{7.19}$$

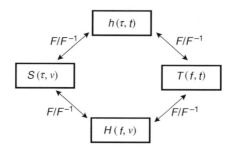

Figure 7.5 Relations between the four deterministic time-varying channel functions

Figure 7.5 shows the interrelationships between the various system functions for characterizing deterministic time-variant linear systems, where F and F^{-1} mean Fourier and inverse Fourier transforms, respectively.

7.4 Stochastic Description of the Wireless Channel

The channel behavior, that we considered in the previous section to be deterministic, in fact, cannot be predicted. This is why *statistical models* are used for characterizing the channel. In a stochastic model, the channel has to be described in terms of *probabilities*. For example, the *impulse response*, $h(t, \tau)$, can be considered as a *random process*. The channel is thus defined by means of the *autocorrelation function* (ACF) of the impulse response, i.e.,

$$R_h(\tau_1, \tau_2, t_1, t_2) = \mathrm{E}[h(\tau_1, t_1)h^*(\tau_2, t_2)] \tag{7.20}$$

where * means complex conjugate. Meanwhile, the mean, $\mathrm{E}[h(\tau, t)]$, is assumed to be zero. This description can be further simplified using the following assumptions [6]:

1. The stochastic process, described by the impulse response, $h(t, \tau)$, is *wide sense stationary*, WSS. In this case the ACF depends only on $\Delta t = t_2 - t_1$, and not on the absolute time instant, t, i.e.,

$$R_h(\tau_1, \tau_2, \Delta t) = \mathrm{E}[h(\tau_1, t)h^*(\tau_2, t + \Delta t)] \tag{7.21}$$

2. The amplitudes and phases of different paths are uncorrelated: *uncorrelated scattering* (US). Then, the ACF of the impulse response disappears for $\tau_1 \neq \tau_2$, and shows a delta-like behavior for $\tau_1 = \tau_2$.

The above assumptions correspond to the so-called WSSUS (*wide sense stationary-uncorrelated scattering*) channel, which has been demonstrated to be a realistic assumption in many radio channels, and is valid in the case of the mobile multipath channel, at least for short sections of the traveled route. This channel can be defined by the autocorrelation function of the impulse response, which can be simplified to the expression [6]

$$R_h(\tau, \Delta t) = \mathrm{E}[h(\tau, t)h^*(\tau, t + \Delta t)] = Q(\Delta t, \tau_1)\delta(\tau_2 - \tau_1) \tag{7.22}$$

The autocorrelation functions for the other system functions can also be used for describing the channel, while still keeping the WSUSS assumption. Thus, the autocorrelation function for T is [6]

$$R_T(f_1, f_2, \Delta t) = R_T(\Delta f, \Delta t) \qquad (7.23)$$

which depends only on the frequency separation, Δf, and not on the absolute frequencies, f_1 and f_2. The following Fourier transform relationship is fulfilled:

$$R_h(\tau, \Delta t) \leftarrow F_\tau \rightarrow R_T(\Delta f, \Delta t) \qquad (7.24)$$

The WSUSS channel can also be described by the ACFs of the two remaining system functions [6], i.e.,

$$R_h(\tau, \Delta t) \leftarrow F_{\Delta t} \rightarrow R_S(\tau, \nu) \text{ and } R_S(\tau, \nu) \leftarrow F_\tau \rightarrow R_H(\Delta f, \nu) \qquad (7.25)$$

The ACF of the *scattering function*, $R_S(\tau, \nu)$, has a special meaning, as it is proportional to the probability with which the multipath contributions arrive with a given delay, τ, and Doppler shift, ν. For convenience, $R_S(\tau, \nu)$ is also frequently denoted by $S(\tau, \nu)$, as in the deterministic case. The relationship between the four ACFs is shown in Figure 7.6.

The above characterization can be further simplified. The ACF of the impulse response, $R_h(\tau, \Delta t)$, calculated for $\Delta t = 0$, and denoted by $p_h(\tau) = R_h(\tau) = R_h(\tau, 0)$, is called the *delay-power spectral density* or *power-delay profile* (PDP). This function describes how the received power is distributed between the different delayed echoes reaching the receiver, and can be converted into a probability density function, $p(\tau)$, if it is normalized to an unit area, i.e.,

$$p(\tau) = \frac{R_h(\tau)}{\int_{-\infty}^{\infty} R_h(\tau)d\tau} = \frac{p_h(\tau)}{\int_{-\infty}^{\infty} p_h(\tau)d\tau} \qquad (7.26)$$

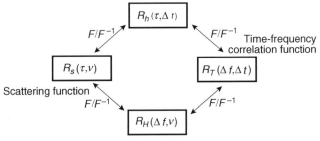

Figure 7.6   Relations between the four ACFs of the random WSSUS channel

The standard deviation of this distribution is given by

$$S_\tau = \sqrt{\int_{-\infty}^{\infty} (\tau - D_\tau)^2 p(\tau) d\tau} \qquad (7.27)$$

and describes the *delay spread* produced by the multipath propagation. In the above equation, D_τ, is calculated using the equation

$$D_\tau = \mathrm{E}[\tau] = \int_{-\infty}^{\infty} \tau p(\tau) d\tau \qquad (7.28)$$

which gives the mean delay.

Likewise, the ACF of the frequency–Doppler function, $R_H(\Delta f, \nu)$ for $\Delta f = 0$ is called the *Doppler–power spectral density*, $R_H(\nu)$ [6], also denoted $S(\nu)$. This function describes the distribution of the echo powers along the Doppler shift, ν, dimension, and can be converted into a probability density function, $p(\nu)$, i.e.,

$$p(\nu) = \frac{R_H(\nu)}{\int_{-\infty}^{\infty} R_H(\nu) d\nu} = \frac{S(\nu)}{\int_{-\infty}^{\infty} S(\nu) d\nu} \qquad (7.29)$$

Again, the standard deviation of this distribution, S_ν, given by

$$S_\nu = \sqrt{\int_{-\infty}^{\infty} (\nu - D_\nu)^2 p(\nu) d\nu} \qquad (7.30)$$

and describes the *Doppler spread* caused by the channel, and where

$$D_\nu = \mathrm{E}[\nu] = \int_{-\infty}^{\infty} \nu p(\nu) d\nu \qquad (7.31)$$

is the mean.

For a complete characterization of a radio channel, we need to know the joint pdf, i.e., $p(\tau, \nu)$, which can be obtained from ACF $R_S(\tau, \nu)$, the scattering function. The pdf for the delay alone, $p(\tau)$, can also be computed integrating the joint pdf to remove the other variable, ν. Similarly, $p(\nu)$ can be calculated through the integration of $p(\tau, \nu)$.

Again, a rough characterization of the channel can be carried out through the ACF of $R_T(\Delta f, \Delta t)$ [6] called the *spaced time-spaced frequency correlation function*. If this

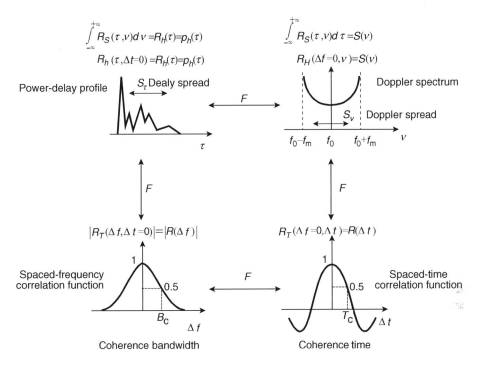

Figure 7.7 Simplified channel characterization functions and parameters

function is particularized for the cases where $\Delta f = 0$, i.e., $R_T(0, \Delta t)$, and $\Delta t = 0$, $R_T(\Delta f, 0)$, i.e., we arrive at the *spaced-time correlation* and *spaced-frequency correlation functions*. These parameters provide a measure of how much the transmission characteristics of the channel vary with the time and frequency spacing. From these correlation functions, the values of the *coherence time*, T_c, and *coherence bandwidth*, B_c, can be calculated. Figure 7.7 provides the overall framework with the above simplified functions.

The *coherence time* is the period, T_c, over which the magnitude of the spaced-time correlation function is at least half its maximum value. During this period, it can be assumed that the transfer function changes only slightly. The *coherence bandwidth* is the bandwidth, B_c, over which the magnitude of the frequency correlation function is larger than half its maximum value. It can be assumed that the transfer function will be almost constant for frequency separations smaller than the coherence bandwidth.

The time spreading and coherence bandwidth will be dependent on the type of environment (urban, suburban, rural, etc.), on the terrain irregularity (flat, hilly, mountainous, etc.), and on the type of cell (mega-, macro-, micro-, picocell). The Doppler spread and coherence time will depend on the *mobility* of the terminal (stationary, pedestrian, vehicular, fast train, etc.). In [7] delay spread values for different propagation scenarios at 900 and 1800/1900 MHz are quoted, e.g., 10–25 µs for urban areas, 200–300 ns for suburban areas, 10–50 ns for indoor scenarios.

For example (`intro71`), assuming we could measure the response to a delta with an ideally infinite delay resolution, and we got the power-delay profile depicted in Figure 7.8(a),

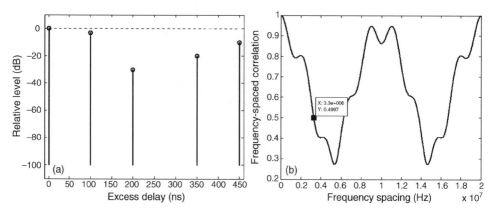

Figure 7.8 (a) Example ideal power-delay profile. (b) Associated frequency-spaced correlation function

it is possible to calculate the PDP parameters which, for this discrete delta case, can be redefined as follows [7],

$$D_\tau = \frac{\sum \tau_i p_h(\tau_i)}{\sum p_h(\tau_i)} \text{ and } S_\tau = \sqrt{\frac{\sum (\tau_i - D_\tau)^2 p_h(\tau_i)}{\sum p_h(\tau_i)}} \tag{7.32}$$

Script `intro71` provides an assessment of the average excess delay and the delay spread in ns, i.e.,

```
D = 61.2948
S = 112.6806.
```

Moreover, the Fourier transform of the PDP has been calculated (Figure 7.8(b)), which provides an assessment of the frequency-spaced correlation function. The transform has been normalized to give a maximum value of one. Reading the frequency separation for a correlation value of 0.5, we can compute the coherence bandwidth. In Figure 7.8(b) we can read a value of 3.3 MHz.

The *coherence time*, T_c, and *coherence bandwidth*, B_c, have a considerable effect on the design of transmission systems. It is possible to distinguish between the following cases when taking into account the *symbol duration*, T_s, and the signal *bandwidth*, $B \approx 1/T_s$ [6] (Figure 7.9):

1. Regarding the symbol duration, T_s:
   (a) If $T_c \ll T_s$, the channel varies during the transmission of a single symbol. Hence, large distortions will result, which will give rise to high BERs. The *Doppler spread*, S_ν, will be greater than the signal bandwidth ($S_\nu \gg B$), and the received signal will show a strong time-variant behavior.
   (b) If $T_c \gg T_s$, the channel can be considered constant during the transmission of one symbol, and the received signal will show a time-invariant behavior.

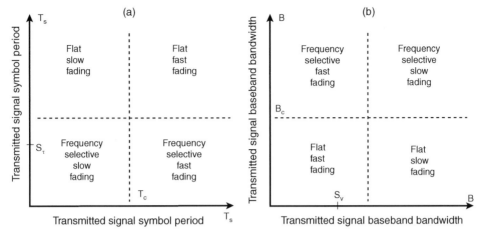

Figure 7.9 Channel classification. (a) With respect to the symbol period. (b) With respect to the baseband signal bandwidth [7]

2. Regarding the transmitted bandwidth, B:
   (a) If $B_c \ll B$, the channel transfer function changes over the signal bandwidth, B. The *delay spread* will be larger than the symbol duration ($S_\tau \gg T_s$), and inter-symbol interference (ISI) will occur. In the frequency domain, a frequency-selective behavior will be observed, with an equalizer being required at the receiver. This behavior is known as *frequency-selective fading*, and we can consider that we have a *wideband channel.*
   (b) If $B_c \gg B$, the transfer function will be approximately constant over the signal bandwidth B. Only small ISI will result. This behavior is known as *non-frequency-selective* or *flat fading*, and we can consider that we have a *narrowband channel.*

7.5 Projects

Next we are going to present several simulation projects that will try and illustrate the concepts introduced above. The first three projects are based on the multiple point-scatterer model. The last deals with the stochastic generation of tapped delay-line time series.

Project 7.1: Time-Varying Frequency Response

Connected with the simulations carried out in Chapter 4, where we used the multiple point-scatterer model for synthesizing the signal's time variations, at or in the neighborhood of the carrier, here we expand the simulator to a wider range of frequencies. This will allow us to explicitly observe the selectivity of the channel (in the frequency domain) at the same time that we view the *time selectivity* (fading) of the channel, as we did in Chapters 4 and 5. Throughout this chapter we consider that the direct LOS contribution is totally blocked, and that the only contributions to the overall received signal are due to multipath.

In `project711` a *two-ray* situation is studied where the difference in delay between the two echoes is small: a small delay spread. We also try to calculate the time-varying impulse response by computing its inverse FFT.

The basic difference between the simulators developed in Chapter 4 and the ones we developed for this chapter reside in the fact that now we also scan in the frequency axis, whereas before we restricted ourselves to a single frequency. The channel time-varying frequency response, in the frame of the multiple point-scatterer model, is given by (Section 7.2)

$$T_{CE}(t,f) = \sum_{i=1}^{N} \tilde{a}_i \exp[-j2\pi(f+f_0)\tau_i(t)] \tag{7.33}$$

where $\tau_i(t) = d_i(t)/c$, with c, the speed of light.

If we study this function only for the carrier frequency, f_0, we arrive at the complex envelope that we studied in Chapters 4 and 5, i.e.,

$$r(t) = T_{CE}(t,f=0) = \sum_{i=1}^{N} \tilde{a}_i \exp[-j2\pi f_0\tau_i(t)] \tag{7.34}$$

where, in the low-pass equivalent domain, the carrier is represented by $f = 0$.

In this project, f, or alternatively, the wavelength, takes on a range of values covering the frequencies of interest, i.e., the transmission bandwidth and adjacent bands. Note that we are treating the channel as deterministic. In fact, we are only generating one of the possible realizations of a random process. Even if it were possible to figure out the actual values of the amplitudes and delays through deterministic propagation models, e.g., using ray tracing, there would always be unpredictable elements such as the phases of each contribution, contained in the complex term $\tilde{a}_i$.

The simulated scenario is shown in Figure 7.10(a), where two scatterers are present. Their magnitudes are 1 and 0.5, respectively. The MS route passes through the origin and BS is

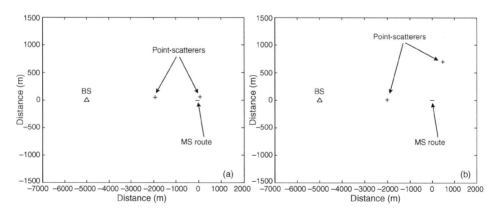

Figure 7.10   (a) Simulated scenario in `project711`. (b) Simulated scenario in `project712`

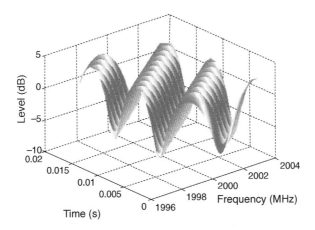

Figure 7.11   Time-varying frequency response

represented by a triangle. We performed the sampling in the traveled distance, with a rate of F=16 samples per wavelength, the mobile speed was 10 m/s, the number of route samples was 20. The band of interest was axis=[1997.5:step_f:2002.5] with a sample spacing in the frequency axis, step_f=0.01 MHz. Figure 7.11 shows the time-varying frequency response, which is equivalent to calculating the complex envelope, $r(t)$, as we did in previous chapters, for all frequencies of interest. We can see in the figure how, in addition to the *time selectivity* (time variations), there is also a *frequency selectivity* effect. Figure 7.12(a) shows the frequency response for the first point of the MS route.

From the time-varying frequency response, $T(t, f)$, we can try to compute the time-varying impulse response, $h(t, \tau)$, by performing the inverse Fourier transform with respect to variable f. Figure 7.13 illustrates the calculated $h(x, \tau)$, where time has been replaced by traveled distance, and Figure 7.12(b) shows $h(\tau)$ for the first point of the MS route.

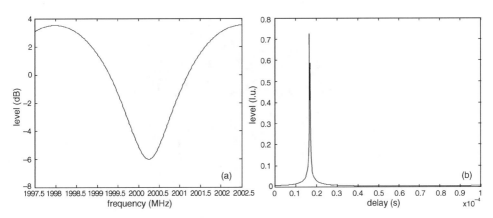

Figure 7.12   (a) Frequency response for the first point of the MS route. (b) Impulse response for the first point of the MS route

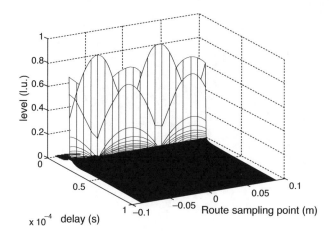

Figure 7.13 Calculated $h(x,\tau)$, where x is the traveled distance

Note in these figures the effect of the limited frequency span in the inverse FFT. The ideal channel is made up of two deltas which are widened, not being resolvable. The available time resolution is given by $\Delta\tau = 1/(f_{max} - f_{min})$. Moreover, we have selected a case where the total propagation time is smaller than the maximum measureable delay, $\tau_{max} = 1/\Delta f$, where Δf is the sampling spacing in the frequency domain. If the propagation distance was longer, the measured delay would be given in modulo-τ_{max} format. The number of inverse FFT points employed was 1024.

In `project712`, we increased the delay difference between the two rays in the simulated two-ray scenario (Figure 7.10(b)). Now, in Figure 7.14 we can see a marked selectivity effect with much more frequent lobes as we scan along the frequency range of interest. This can be

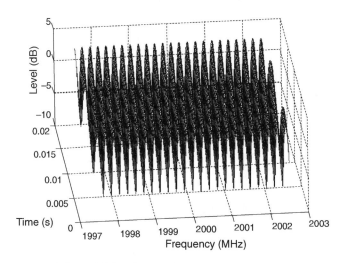

Figure 7.14 Time-varying frequency response

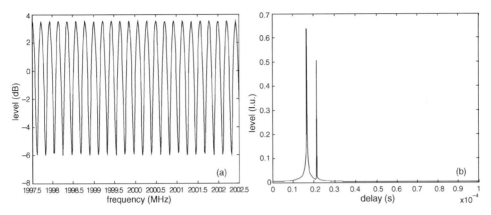

Figure 7.15 (a) Frequency response for first point of MS route. (b) Impulse response for first point of MS route

seen in a clearer way in Figure 7.15(a), where we illustrate the frequency response for the first point of the MS route. In Figure 7.16 we see a sequence of impulse responses estimated by means of the inverse Fourier transform. In this case, the time separation between the two echoes is greater than the available resolution, and we can clearly tell one echo from the other. This is also shown in Figure 7.15(b) for the first point of the MS route.

In `project713` we have simulated the scenario depicted in Figure 7.17. This is a more complex, but more realistic channel, and the obtained results also look more realistic. We would like to point out some of the modifications introduced with respect to the previous two projects. In this case, we specified the overall relative power with respect to a reference, i.e., free-space conditions. To get the link budget right, we need to select the magnitudes of the point scatterers accordingly. This has already been discussed in Chapter 5. There, however, we considered all magnitudes to be equal. Here, we introduce a distance decay law that

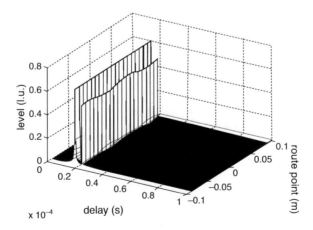

Figure 7.16 Sequence of time-varying impulse responses

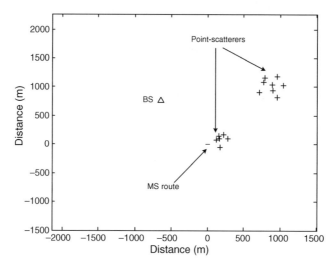

Figure 7.17 Simulated scenario in `project713`

should provide more realistic results. We have assumed that we have isotropic point scatterers with the same radar cross-section, thus, the only differences in the scatterer magnitudes will be due to the distance.

If we normalize with respect to the unblocked direct signal, the relative amplitude of the contribution from a given point scatterer, with sub-path distances d_{i-1} and d_{i-2}, is

$$a_i = \frac{\dfrac{\sqrt{\sigma_i}}{d_{i-1}d_{i-2}}}{\dfrac{1}{d_0}} \tag{7.35}$$

where we have calculated the ratio between the magnitudes of the scattered and the direct signal, and where d_0 is the direct signal radio path. If we want to specify a given total normalized relative power, $\bar{P}'$ (dB), the following relationship must be fulfilled,

$$\sum_i a_i^2 = 10^{\bar{P}'/10} \tag{7.36}$$

One interesting thing to point out is that if we need to generate larger cross-section scatterers, we can introduce several at approximately the same position. This brings up the concept of *echo clustering* that we will be discussing in Chapters 8 and 10.

We can see in Figures 7.18–7.21 how both the time-varying frequency response and the time-varying impulse response include much more realistic features than in the case of the simple two-ray model.

Finally, script `project714` is provided which contains a scenario editor. When the screen showing BS and the MS route is presented to the user, point scatterers can be input graphically by pointing the cursor with the mouse to the wanted location, and clicking on the

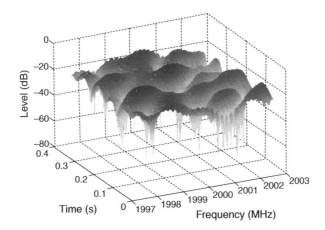

Figure 7.18   Time-varying frequency response

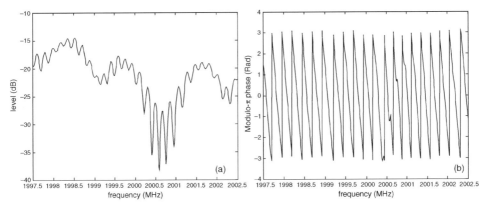

Figure 7.19   (a) Frequency response for first point of MS route. (b) Phase response for first point of MS route

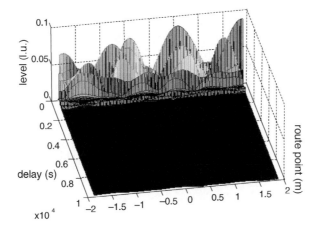

Figure 7.20   Time-varying impulse response

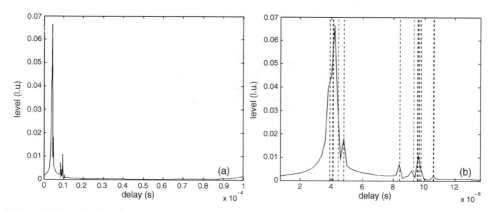

Figure 7.21 (a) Impulse response for first point of MS route. (b) A closer look at the impulse response for first point of MS route and in broken lines the nominal delays for the point scatterers

left button. The scenario editor is exited by clicking on the right button. This last click introduces one extra point scatterer. The editor can also be exited by pressing ENTER.

Project 7.2: Ideal Channel in the Time-Delay Domain and Channel Sounding

In wideband channel studies, one of the main pieces of information sought is the power-delay profile and its associated parameters: the delay spread and the coherence bandwidth, obtained from the Fourier transform of the PDP. Measurements are normally carried out in the time domain. Sounding may also be carried out in the frequency domain by scanning the measurement bandwidth for obtaining an estimate of the channel transfer function, $T(f, t)$. This is mainly found in indoor scenarios where vector analyzers are used. The limitation comes from the need that the transmit and receive antennas have to be hooked up to the instrument by means of not too long cables.

Two possible methods for measuring power-delay profiles in the time domain are: (a) using regular pulses and (b) using pulse compression techniques [2].

With the first method, high peak powers are required, and this produces interference problems. Generally, *pulse compression techniques* are used. These are based on generating sounding signals with similar characteristics to those of white noise. If white noise, $n(t)$, is applied to the input of a linear system, and we compute the cross-correlation of the received signal, $w(t)$, with a replica of the transmitted noise shifted a time τ, $n(t - \tau)$, we obtain $h(\tau)$. This is based on the important feature of white noise whose autocorrelation is a delta, i.e.,

$$E[n(t)n^*(t - \tau)] = R_n(\tau) = N_0 \, \delta(\tau) \tag{7.37}$$

where $R_n(\tau)$ *is* the autocorrelation function of $n(t)$, and N_0 is its one-sided power spectral density. The output of the measurement system (*channel sounder*) is given by

$$w(t) = \int_{-\infty}^{\infty} h(\xi)n(t - \xi)d\xi \tag{7.38}$$

Computing the cross-correlation between the input and the output, yields [2]

$$E[n(t)n^*(t - \tau)] = E\left[\int_{-\infty}^{\infty} h(\xi)n(t - \xi)n^*(t - \tau)d\xi\right]$$

$$= \int_{-\infty}^{\infty} h(\xi)R_h(\tau - \xi)d\xi = N_0\,h(\tau) \tag{7.39}$$

The result is thus a sample of the channel impulse response for a specific delay, τ. Hence, by scanning through all values of τ of interest, we can obtain the complete impulse response. In practice, deterministic signals are used instead of noise, which present similar character-istics to those of white noise. Such sequences are known as *pseudo-noise sequences*, PN, of length m. These present autocorrelation characteristics similar to a delta (Figure 7.22). In fact, it is a narrow triangular pulse with a base width equal to $2\tau_0$, where τ_0 is the PN sequence clock period. The sounding process requires that the clocks from which the PN sequences at the transmitter and receiver are generated be synchronized.

Effectively, the *channel sounding* operation approximately produces the convolution between the actual ideal channel (made up of deltas) and the sounding impulse. Each sounding cycle produces an estimation of the *instantaneous power-delay profile* (PDP), i.e.,

$$p_h(\tau, t) = |h(\tau, t)^* h_{\text{ChannelSounder}}(\tau)|^2 \tag{7.40}$$

To prevent fade effects from affecting the measurement, it is necessary to average several consecutive instantaneous profiles. In this way, possible fades or enhancements showing on the instantaneous profiles, at given delays, are smoothed out. Thus, the average PDP (APDP) is defined as

$$p_h(\tau) = \frac{1}{N}\sum_{n}^{N} p_h(\tau, t) \tag{7.41}$$

which is normally put in dB, i.e., $P_h(\tau) = 10\log[p_h(\tau)]$.

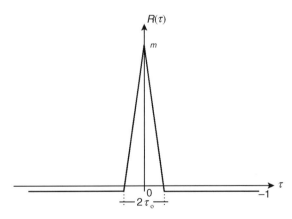

Figure 7.22   Autocorrelation function of a pseudo noise sounding sequence

The characterization of the wideband channel is normally carried out in two stages, as for the narrowband case: a *small-scale* characterization for short runs of approximately 20λ to 40λ (small area), and a *large-scale* characterization for larger areas.

For the small-scale characterization, from the APDP representing a given local area, two parameters are computed: the *average delay*, D_τ, and the *delay spread*, S_τ, which have already been defined. In the frequency domain, the most important parameter is the *coherence bandwidth*, which is computed from the frequency-spaced correlation function which is related to the power-delay profile and is given by (Figure 7.7)

$$R_T(\Delta f) = \int_{-\infty}^{\infty} p_h(\tau) \exp(-j2\pi\Delta f\tau)d\tau \qquad (7.42)$$

The coherence bandwidth is taken for a frequency separation corresponding to half the maximum of $R_T(\Delta f)$.

For characterizing larger areas, the distributions of the delay spread and coherence bandwidth are computed for the different possible operational environments, i.e., urban, suburban, open, flat, hilly, etc.

We have implemented a simulator in script `project721` where we worked from the channel's ideal time-varying impulse response made up of deltas and we tried to simulate the sounding process in a simplified way. Figure 7.23 shows the simulated scenario. The simulator based on the multiple point-scatterer model generates ideal impulse responses for each point of the MS route as illustrated in Figure 7.24(a). We discretized the delays to a sufficiently small step of 1 ns for ensuing discrete operations.

We simulated the sounding process by performing the convolution of the ideal impulse responses with a triangular pulse where we assumed a sequence clock period, τ_0, of 100 ns (10 MHz) and a sequence length $m = 511$. This pulse was normalized to have unit energy in

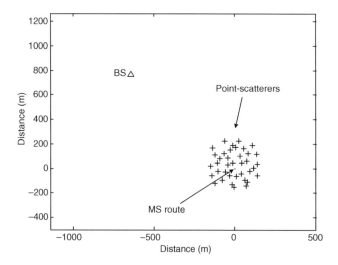

Figure 7.23 Simulated scenario in `project721`

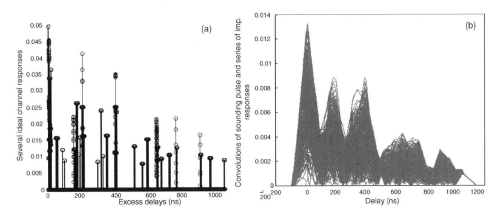

Figure 7.24 (a) Superposed ideal impulse responses for all MS route points. (b) Convolution of all ideal impulse responses with sounding pulse

order not to disrupt the link budget. Figure 7.24(b) shows, superposed, the results of the instantaneous convolutions, and Figure 7.25 shows the sequence of instantaneous PDPs in dB. Finally, Figure 7.26 illustrates the averaged PDP. The process described allows us to make quite realistic simulations, very similar to those found in the literature.

Also, the simulator implements the same procedure as in `intro71` for calculating PDP parameters D_τ and S_τ. We have carried out these calculations twice – one for the ideal power-delay profile (squared magnitude of the ideal impulse response made up of deltas) and the second for the PDP from the simulated sounding process. The results are listed below. For the ideal PDP,

```
D = 317.2284
S = 285.1015
```

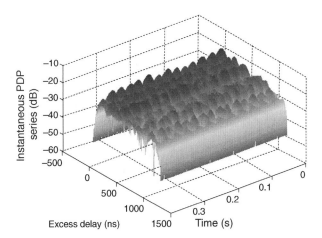

Figure 7.25 Sequence of instantaneous PDPs

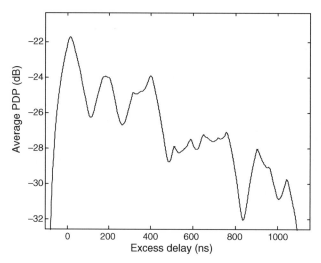

Figure 7.26   Average PDP

and for the simulated PDP after sounding

```
D = 457.4476
S = 312.2812
```

The differences are mainly due to the effect of the finite width of the sounding pulse.

Finally, script `project722` is provided, which implements an scenario editor similar to that in `project714`.

Project 7.3: The Scattering Function

In `project731` we repeat some of the above results, but we try to go one step further and show other channel functions, more specifically, the scattering function, $S(\tau,\nu)$. So far, we have presented the impulse response, $h(\tau,t)$, in two ways, the ideal one made up of deltas, and the one resulting from the inverse transform of $T(f,t)$. We leave the reader to calculate and plot function $H(f,\nu)$. This is a difficult function to visualize and understand since, as we saw earlier in this chapter (Figure 7.4), it represents the channel as a bank of filters with transfer functions, $H(f,\nu)\Delta\nu$, followed by Doppler shifting frequency converters producing Doppler shifts in the range $(\nu, \nu + \Delta\nu)$ Hz.

Here, we try to extend the chain of Fourier transforms to reach to the channel's scattering function, $S(\tau,\nu)$. The settings of the simulator are the same as in previous projects. The input scenario is shown in Figure 7.27. Figure 7.28 shows, superposed, all ideal impulse responses for the MS route. We have again discretized the delay axis to a step of 0.1 μs. Figures 7.29 and 7.30 show the scattering function $S(\tau,\nu)$ both in linear units and dB, computed as the Fourier transform of the time-varying impulse response, $h(\tau,t)$, with respect to the time variable, t, to arrive at the Doppler variable, ν.

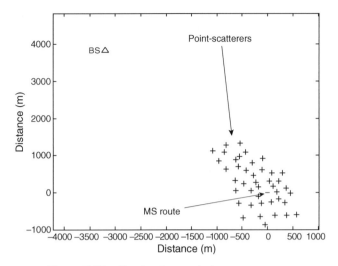

Figure 7.27 Simulated scenario in `project731`

Note the differences in the scattering functions in Figures 7.29 and 7.30 which correspond to the same channel. Their differences reside in the way they have been calculated, which involve one and two Fourier transforms, respectively. Thus, the scattering function in Figure 7.29 was computed following the route shown in the equation below, i.e.,

$$h_{\text{ideal}}(\tau, t) \rightarrow F \rightarrow S(\tau, \nu) \tag{7.43}$$

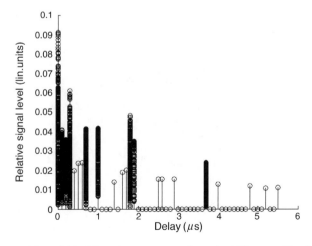

Figure 7.28 Superposition of ideal impulse responses along the MS route. The delay axis has been discretized to 0.1 μs

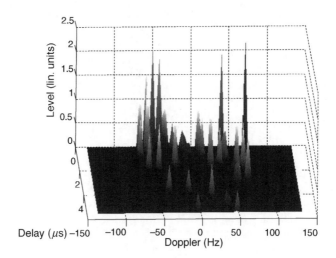

Figure 7.29 Scattering function in linear units resulting from the Fourier transform of the time-varying impulse response

where the start-up function was $h_{\mathrm{ideal}}(\tau, t)$, whereas for computing the scattering function in Figure 7.30 we started from $T(f, t)$, thus,

$$T(f,t) \rightarrow F^{-1} \rightarrow h(\tau,t) \rightarrow F \rightarrow S(\tau, v) \qquad (7.44)$$

Finally, script `project732` is provided which implements an scenario editor similar to that in `project714` and `project722`.

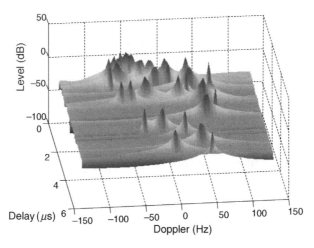

Figure 7.30 Scattering function in dB resulting from the Fourier transform of the time-varying impulse response

Project 7.4: Tapped Delay-Line Models – COST 207

Both the channel and the transmitted signal must be represented in the discrete-time domain in order to carry out simulations. The channel effects can be introduced in this discrete-time framework using a *tapped delay-line* (TDL) model. We have already presented this concept earlier in this chapter (Section 7.3) but considered differential tap spacings.

The sampling rate, f_s, must account for the maximum Doppler, and take into account the signal bandwidth, B. Thus, the required sampling frequency is $f_s \geq 2(B + f_D)$. The discrete-time channel model assumed is a transversal or FIR filter with time-varying taps, $g_i(t)$, spaced $\Delta\tau = t_s$ (Figure 7.31).

If the WSSUS assumption is made (Section 7.4), the complex coefficients, $g_i(t)$, are uncorrelated, and Gaussian distributed (central limit theorem). This model is a direct representation of the joint delay-Doppler probability distribution, $p(\tau, \nu)$, which is proportional to the scattering function, $R_S(\tau, \nu)$. In the COST 207 model [8] this joint probability density function is given in the form

$$p(\tau, \nu) = p(\tau)p(\nu|\tau) \tag{7.45}$$

i.e., the channel is modeled in terms of the power-delay distribution, $p(\tau)$, and of the Doppler distribution conditioned to each specific delay, $p(\nu|\tau)$.

Looking at Figure 7.31, we can see two elements in each tap, R_i and W_i: the first is a normalized complex Rayleigh distribution with a given Doppler spectrum, and the second term is a weight indicating the power share for that specific tap. Combining these two terms into a single, time-varying coefficient, $g_i(t)$, the corresponding Rayleigh distribution parameter is given by

$$2\sigma_{g_i}^2 = \int_{it_s}^{(i+1)t_s} p(\tau)d\tau \tag{7.46}$$

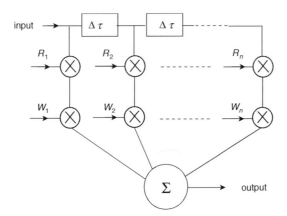

Figure 7.31   TDL model representing a time-varying channel

the tap spacing and the time sample spacing being $t_s = 1/f_s$. Parameter σ_{g_i} is the standard deviation for the two Gaussian random number generators in quadrature used for generating the complex samples of each tap, $g_i(t)$, as we discussed in Project 5.4.

One further simplification, not made in the case of the COST 207 model implemented in project74, is assuming that the scattering function $R_S(\tau, \nu)$ and, consequently, also the joint PDF, $p(\tau, \nu)$, is *separable*. This assumption is equivalent to the statistic independence of both random variables, τ and ν, i.e.,

$$p(\tau, \nu) = p(\tau)p(\nu) \tag{7.47}$$

This assumption would simplify considerably the simulation process.

In project COST 207 [8], several wideband propagation models were proposed for the practical realization of both hardware and software simulators in the context of GSM systems. These models are based on the following transversal filter elements (Figure 7.31):

1. A number of taps, each one with its corresponding delay and average power. In this model, unevenly spaced taps were proposed (Table 7.1).
2. A complex Rayleigh time series for each tap, with an associated Doppler spectrum $S(\tau_i, \nu)$ or, equivalently $p(\nu|\tau_i)$, where i indicates the tap number (Figures 7.32(a) and (b)).

Typically, the assumed power-delay pdf is an exponential distribution, i.e.,

$$p(\tau) = a \exp(-\tau/b) \tag{7.48}$$

or a combination of two exponentials for long delay scenarios, corresponding to irregular terrain or mountainous areas. Exponential distributions become linear when the power is put in dB. The proposed power-delay distributions are summarized in Figure 7.32(a).

Table 7.1 COST 207 model parameters for the bad urban (BU) scenario [8]

Tap No.	Delay [μs]	Power (lin)	Power [dB]	Doppler category
1	0	0.2	−7	CLASS
2	0.2	0.5	−3	CLASS
3	0.4	0.79	−1	CLASS
4	0.8	1	0	GAUS1
5	1.6	0.63	−2	GAUS1
6	2.2	0.25	−6	GAUS2
7	3.2	0.2	−7	GAUS2
8	5.0	0.79	−1	GAUS2
9	6.0	0.63	−2	GAUS2
10	7.2	0.2	−7	GAUS2
11	8.2	0.1	−10	GAUS2
12	10.0	0.03	−15	GAUS2

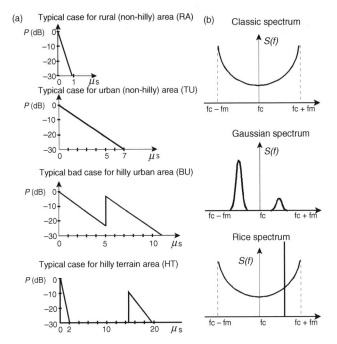

Figure 7.32   COST 207 model. (a) Power-delay profiles. (b) Doppler spectra

Four types of Doppler spectra were defined conditioned to their associated delays (Figure 7.32(b)), i.e.,

(a) *Classic spectrum* (CLASS), used for delays not greater than 500 ns, and given by

$$S(\tau_i; f) = \frac{A}{\sqrt{1 - (f/f_{\mathrm{m}})^2}} \quad \text{for } f \in (-f_{\mathrm{m}}, f_{\mathrm{m}}) \tag{7.49}$$

f_{m} being the maximum Doppler shift, and where A can be used to normalize the filter, in order not to distort the tap average power. This corresponds to the Clarke/Jakes Doppler spectrum discussed in Chapter 5.

(b) *Gaussian spectrum 1* (GAUS1), given by the sum of two Gaussian functions, and associated to delays in the range between 500 ns and 2 μs

$$S(\tau_i; f) = G(A, -0.8f_{\mathrm{m}}, 0.05f_{\mathrm{m}}) + G(A_1, 0.4f_{\mathrm{m}}, 0.1f_{\mathrm{m}}) \tag{7.50}$$

where A_1 is 10 dB below A, with

$$G(A, f_1, f_2) = A \exp\left[-\frac{(f - f_1)^2}{2f_2^2}\right] \tag{7.51}$$

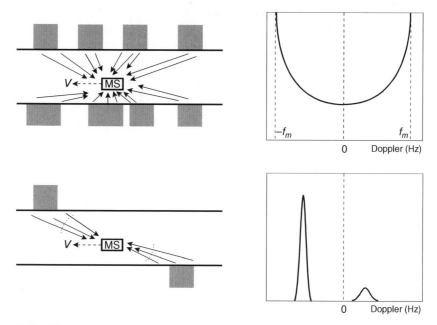

Figure 7.33 Classic and Gaussian Doppler spectra, and schematic representation form the point of view of the angles of arrival of the multipath echoes

Figure 7.33 illustrates the spatial distribution of echoes behind both the classic (Clarke/Jakes, Chapter 5) and Gaussian Doppler spectra, assumed throughout this model.

(c) *Gaussian spectrum 2* (GAUS 2), given by the sum of two Gaussian functions, and associated to paths with delays in excess of 2 μs

$$S(\tau_i;f) = G(B, -0.7f_m, 0.1f_m) + G(B_1, 0.4f_m, 0.15f_m) \tag{7.52}$$

where B_1 is 15 dB below B.

(d) *Rice spectrum* (RICE), given by the sum of a classic Doppler spectrum and a delta representing the direct ray, i.e.,

$$S(\tau_i;f) = \frac{0.41}{2\pi f_m \sqrt{1 - (f/f_m)^2}} + 0.91\delta(f - 0.7f_m) \text{ for } f \in (-f_m, f_m) \tag{7.53}$$

This spectrum is used for the shortest path/tap for the rural area case. Figure 7.32(b) shows all these spectra.

As an example, in Table 7.1 the specifications for the so-called *bad urban* (BU) scenario are summarized, where 12 taps were defined each with a given delay, weight and spectral shaping. This channel has been implemented in `project74`.

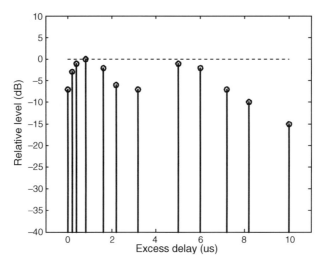

Figure 7.34   Relative powers for all 12 taps

Figure 7.34 illustrates the distribution of powers associated to each the tap. Figures 7.35(a) and (b) show the definition of the Gaussian filter in the frequency and time domains. The procedure followed for simulating the classic Doppler filter has already been discussed in Project 6.1. The Gaussian filter, in the discrete-time domain is a sampled version of the continuous filter representation where the transform pair below was used,

$$\sqrt{2\pi}\, f_2 \exp(-2\pi^2\, f_2^2 t^2) \exp(\mathrm{j}2\pi f_1 t) \overset{F}{\longleftrightarrow} \exp\left[-\frac{(f-f_1)^2}{2f_2^2}\right] \qquad (7.54)$$

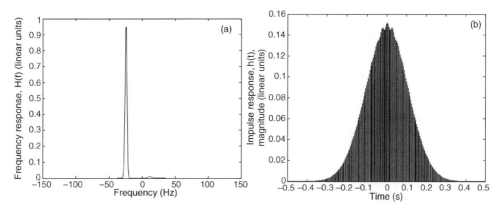

Figure 7.35   (a) Frequency domain representation of one of the Doppler Gaussian filters. (b) Discrete time-domain representation of one of the Doppler Gaussian filters

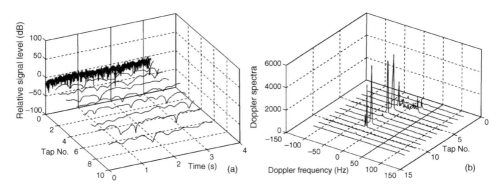

Figure 7.36   (a) Time series for all 12 taps. (b) Doppler spectra for all 12 taps

Finally, Figures 7.36(a) and (b) illustrate the time series of the magnitudes and corresponding Doppler spectra for all the taps.

7.6 Summary

In our step-by-step study, we have introduced concepts relative to the wideband channel. When the time spreading introduced by the channel is significant with respect to the duration of the transmitted symbols, selective fading starts to be meaningful. We have presented simple techniques for characterizing wideband channels, starting with deterministic time-varying channels, and then going on to introduce the stochastic characterization. The main functions and parameters necessary have been discussed. We have put the wideband channel in the frame of the multiple point-scatterer model, which has been used for presenting illustrative channel simulations. Finally, a statistical model, that of COST 207, has been implemented.

References

[1] P.A. Bello. Characterization of randomly time-variant linear channels. *IEEE Trans. Comm.*, **11**, 1963, 360–390.

[2] J.D. Parsons. *Mobile Radio Propagation Channel*, second edition, John Wiley & Sons, Ltd, Chichester, UK, 2000.

[3] F.P. Fontan, M.A.V. Castro & P. Baptista. A simple numerical propagation model for non-urban mobile applications. *IEE Electron. Lett.*, **31**(25), 1995, 2212–2213.

[4] J.M. Hernando & F. Pérez-Fontán. *An Introduction to Mobile Communications Engineering*, Artech House. Boston, 1999.

[5] J.K. Cavers. *Mobile Channel Characteristics*, Kluwer Academic, 2000.

[6] H. Rohling. Lecture Mobile Communications. June 2000, www.et2.tu-harburg.de/Mitarbeiter/Rohling/.

[7] T.S. Rappaport. *Wireless Communications. Principles and Practice*, Prentice Hall, 1996.

[8] Digital land mobile radio communications. Cost 207. Final Report. 1989. Commission of European Communities, Brussels.

Software Supplied

In this section we provide a list of functions and scripts developed in MATLAB®
(MATLAB® is a registered trademark of The MathWorks, Inc.) implementing the various
projects and theoretical introductions mentioned in this chapter. They are the following:

```
intro71                    gaussCOST207
project711                 jakes
project712                 rayleighCDF
project713                 stem2D
project714
project721
project722
project731
project732
project74
```

8

Propagation in Microcells and Picocells

8.1 Introduction

With the massive growth of wireless systems, short-range cells are needed in order to provide an adequate signal to an increasing number of users with a suitable *quality of service* (QoS). This has led to the deployment of *microcells* (cells with base stations below average rooftop heights, Figure 8.1) and *indoor picocells*. Also, indoor coverage from outdoor BSs is a common feature in current wireless systems. Even indoor-to-outdoor signals from, for example, WLAN access points can be strong, thus polluting the RF environment. In this chapter we analyze and simulate some of the basic issues relative to these channels.

8.2 Review of Some Propagation Basics

In this chapter, we will be performing some simple ray tracing involving reflected rays. Thus, we would like to quantify the *reflection* phenomenon. Additionally, the indoor scenario is characterized by the effect of walls blocking the direct path. Hence, we also will be studying the *transmission* phenomenon. The other mechanism used in this chapter, diffraction, was dealt with extensively in Chapter 2.

When a radiowave impinges on a flat surface obstruction with different material parameters than the propagation medium, and dimensions greater than the wavelength, reflection and transmission will occur. In the case of smooth homogeneous interfaces, the wave interaction follows *Snell's laws of reflection and refraction* (Figure 8.2). The law of reflection states that the angle of the reflected field, θ_r, is equal to the angle of the incident field, θ_i, where these angles are referred to the surface's normal. Snell's law of refraction states that the angle, θ_t, is a function of the incidence angle and the materials of the two media, and is given by the following formula,

$$\frac{\sin(\theta_i)}{\sin(\theta_t)} = \sqrt{\frac{\varepsilon_2 \mu_2}{\varepsilon_1 \mu_1}} = \frac{n_2}{n_1} \tag{8.1}$$

Modeling the Wireless Propagation Channel   F. Pérez Fontán and P. Mariño Espiñeira
© 2008 John Wiley & Sons, Ltd

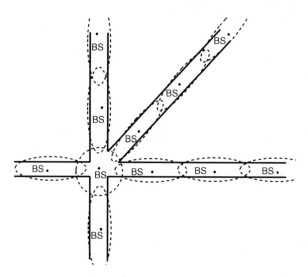

Figure 8.1 Schematic diagram of a microcell scenario in an urban area. Microcells can cover ranges of several hundred meters

where n_1 and n_2 are the refraction indexes for both media, ε_1 and ε_2 their permittivity values, and μ_1 and μ_2 their permeability values. The refraction index of a medium is the ratio of the free-space velocity, c, to the phase velocity of the wave in that medium, i.e.,

$$n = \frac{c}{v} = \sqrt{\frac{\varepsilon\mu}{\varepsilon_0\mu_0}} = \sqrt{\varepsilon_r\mu_r} \tag{8.2}$$

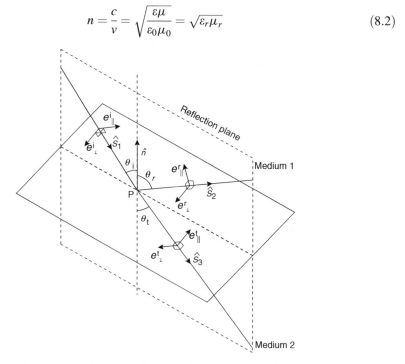

Figure 8.2 Polarizations in transmissions and reflections

The β_1 and β_2 parameters, the phase constants for both media, are related to the refraction indexes through the angular frequency, ω, i.e.,

$$\beta = \omega\sqrt{\mu\varepsilon} = \omega\sqrt{\mu_0\varepsilon_0}\sqrt{\mu_r\varepsilon_r} = \omega\frac{n}{c} \tag{8.3}$$

Moreover, the incident ray, the normal vector to the interface at the refraction point and the refracted ray are on the same plane.

To calculate the reflected and the transmitted fields, the *Fresnel coefficients of reflection and transmission* can be used. These coefficients depend on the constitutive parameters of the materials, the polarization of the incident field and the angle of incidence. The coefficients are different for the *parallel* ($\parallel$) and *perpendicular* ($\perp$) *polarization*. Figure 8.2 illustrates the definition of the parallel and perpendicular electric fields, which are taken with respect to the plane containing the incident, reflected and transmitted rays. The parallel components are contained in the reflection plane while the perpendicular components are perpendicular to the reflection plane.

For reflections on non-perfectly conducting surfaces, the plane wave Fresnel reflection coefficients are given by

$$R_\perp(\theta) = \frac{\cos(\theta) - \sqrt{\varepsilon' - \sin^2(\theta)}}{\cos(\theta) + \sqrt{\varepsilon' - \sin^2(\theta)}} \tag{8.4}$$

and

$$R_\parallel(\theta) = \frac{\varepsilon'\cos(\theta) - \sqrt{\varepsilon' - \sin^2(\theta)}}{\varepsilon'\cos(\theta) + \sqrt{\varepsilon' - \sin^2(\theta)}} \tag{8.5}$$

where $\varepsilon' = \varepsilon/\varepsilon_0 - j\sigma/(\omega\varepsilon_0)$, with $\varepsilon = \varepsilon_r\varepsilon_0$ the permittivity of the material, σ the conductivity, and θ the angle of incidence formed by the incident ray and the normal vector to the surface.

The incident field can be expressed in terms of its parallel and perpendicular components, i.e., $\vec{e}_i = e_{i\parallel}\hat{a}_\parallel + e_{i\perp}\hat{a}_\perp$. The reflected field will also be given in terms of these two components, and their respective reflection coefficients, i.e.,

$$\vec{e}_r = e_{r\parallel}\hat{a}_\parallel + e_{r\perp}\hat{a}_\perp = e_{i\parallel}R_\parallel\hat{a}_\parallel + e_{i\perp}R_\perp\hat{a}_\perp \tag{8.6}$$

The transmitted field can be expressed in the same way. The transmission coefficients are given by the formulas

$$T_\perp = \frac{2\cos(\theta_i)}{\cos(\theta_i) + \sqrt{\varepsilon' - \sin^2(\theta_i)}} \tag{8.7}$$

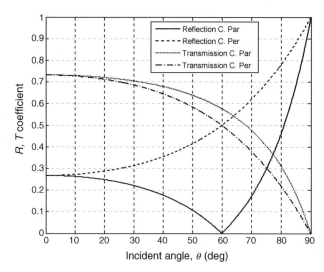

Figure 8.3 Magnitude of reflection and transmission coefficients for very dry ground at 2 GHz

and

$$T_\| = \frac{2\sqrt{\varepsilon'} \cos\left(\theta_i\right)}{\varepsilon' \cos\left(\theta_i\right) + \sqrt{\varepsilon' - \sin^2\left(\theta_i\right)}} \tag{8.8}$$

In Figure 8.3 we have plotted the reflection and transmission coefficients for very dry ground at 2 GHz [1]. Figure 8.4 does the same for wet ground [1] and Figure 8.5 for a brick wall [2].

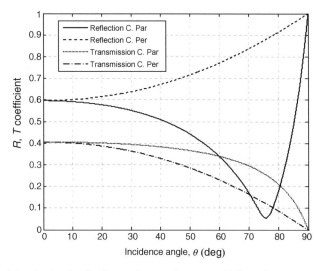

Figure 8.4 Magnitude of reflection and transmission coefficients for wet ground at 2 GHz

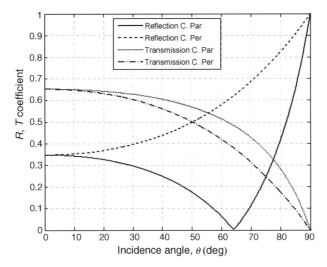

Figure 8.5 Magnitude of reflection and transmission coefficients for a brick wall at 2 GHz

Note how, for propagation almost parallel to the ground, the parallel coefficients can be identified with the *vertical polarization coefficients* and the perpendicular polarization coefficients with the *horizontal polarization coefficients*. However, for wall reflections this equivalence is reversed. Also note that for perfect conductors, the reflection coefficients take on the values $R_\parallel = +1$ and $R_\perp = -1$, and the transmission coefficients take a null value.

Script `intro81` was used for creating the first two figures and `intro82` the third. Note how the parallel reflection coefficient becomes zero or close to zero at the so-called *pseudo-Brewster angle*, $\theta_B = \tan^{-1}(n_2/n_1)$. The reader is encouraged to modify these scripts in order to plot the phases for the reflection coefficients. Constants for other media or materials can be found in Table 8.1 [2]. Try other materials and also the effect of the conductivity.

Table 8.1 Material properties [2]

Material	ε'_r	σ (S/m)	ε''_r @ Frequency
Glass	3.8–8		$< 3 \times 10^{-3}$ @ 3 GHz
Wood	1.5–2.1		< 0.07 @ 3 GHz
Gypsum board	2.8		0.046 @ 60 GHz
Chip board	2.9		0.16 @ 60 GHz
Dry brick	4		0.05–0.1 @ 4, 3 GHz
Dry concrete	4–6		0.1–0.3 @ 3, 60 GHz
Aerated concrete	2–3		0.1–0.5 @ 3, 60 GHz
Limestone	7.5	0.03	
Marble	11.6		0.078 @ 60 GHz
Ground	7–30	0.001–0.030	
Fresh water	81	0.01	
Seawater	81	4	
Snow	1.2–1.5		$< 6 \times 10^{-3}$ @ 3GHz
Ice	3.2		2.9×10^{-3} @ 3GHz

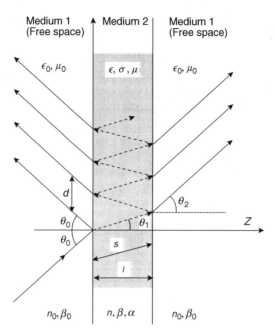

Figure 8.6 Wall reflection, loss and transmission

In the context of indoor, or outdoor-to-indoor propagation, the characterization of wall attenuation is important. Here, a common model for walls is presented based on a slab of finite depth and uniform structure as illustrated in Figure 8.6. The following assumptions can be made with regard to transmissions through walls: the wall medium is homogeneous and isotropic, and the two interfaces are locally plane at the transmission points. Applying Snell's laws, it can be observed how the incidence angle on the first air–wall interface is equal to the exit angle on the second interface, i.e., wall–air. Thus, for the first interface,

$$\beta_0 \sin(\theta_0) = \beta \sin(\theta_1) \tag{8.9}$$

and for the second interface,

$$\beta \sin(\theta_1) = \beta_0 \sin(\theta_2) \tag{8.10}$$

Equating both expressions yields $\theta_0 = \theta_2$, which means that the transmitted ray exiting the wall on the other side is parallel to the incoming ray. The transmitted signal is a combination of several rays, as illustrated in Figure 8.6. Also in the figure, the reflected ray is a combination of rays, thus the term *generalized* or *effective reflection* and *transmission coefficient* used below.

In the case of in-building propagation (transmissions through walls), the finite thickness of the surface has to be taken into account. If the thickness is small compared to the wavelength or if the material presents a small loss, a plane wave model based on a succession of reflections within a slab (wall) leads to a fairly realistic formulation of the problem.

Assuming a single slab of, in general, lossy dielectric material, the total reflected field is given by

$$\vec{e}^{\mathrm{r}} = \vec{e}^{\mathrm{i}} \left\{ R + \sum_{n=1}^{\infty} \exp(-2ns\alpha)\exp(-2njs\beta)(1+R)(1-R)(-R)^{2n-1}\exp[jk_0 dn\sin(\theta)] \right\} \quad (8.11)$$

with n the number of double internal reflections and k_0 denotes the free-space phase constant, while α and β are the plane wave attenuation and phase constant of the medium.

The corresponding *effective* or *generalized reflection coefficient*, which is the sum of the multiple reflections within a wall, is given by

$$R_{\mathrm{eff}} = R \left\{ 1 - \frac{(1-R^2)\exp(-2\alpha s)\exp(-2j\beta s)\exp[jk_0 d \sin(\theta)]}{1-R^2\exp(-2\alpha s)\exp(-2j\beta s)\exp[jk_0 d \sin(\theta)]} \right\} \quad (8.12)$$

The same calculations can be carried out for the *effective transmission coefficient*, which includes the effect of the wave transmitted through the wall, and the energy that undergoes several reflections within the wall, i.e.,

$$T_{\mathrm{eff}} = \frac{(1-R^2)\exp(-\alpha s)\exp(-j\beta s)}{1-R^2\exp(-2\alpha s)\exp(-2j\beta s)\exp[jk_0 d \sin(\theta)]} \quad (8.13)$$

The expression for parameters α and β corresponding to a lossy medium are

$$\alpha = \omega\sqrt{\frac{\mu\varepsilon}{2}}\sqrt{\sqrt{1+\left(\frac{\sigma}{\omega\varepsilon}\right)^2}-1} \quad (8.14)$$

and

$$\beta = \omega\sqrt{\frac{\mu\varepsilon}{2}}\sqrt{\sqrt{1+\left(\frac{\sigma}{\omega\varepsilon}\right)^2}+1} \quad (8.15)$$

The parameter s, which is the length of the trajectory traveled by the wave from one to the other facet of the wall (where l is the thickness of the wall), and d, (see Figure 8.6), can be computed as follows:

$$s = \frac{l}{\sqrt{1-\dfrac{\sin^2(\theta)}{\varepsilon_{\mathrm{r}}}}} \quad (8.16)$$

and

$$d = \frac{2l}{\sqrt{\dfrac{\varepsilon_{\mathrm{r}}^2}{\sin^2(\theta)}-1}} \quad (8.17)$$

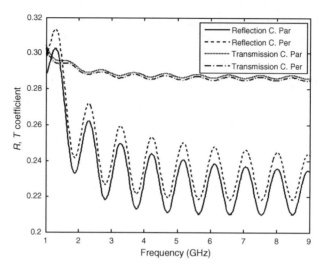

Figure 8.7 Generalized reflection and transmission coefficients as a function of frequency for an incidence angle of $10°$ with $\varepsilon = 2.5\varepsilon_0$, $\sigma = 0.01$ and width $= 0.1$ m

Figures 8.7 and 8.8 illustrate the dependence of the reflection coefficient with the frequency and the slab thickness. These figures were obtained with script `intro83`. The reader is encouraged to try other materials (Table 8.1) and incidence angles. Note the order of magnitude of the transmission loss in Figure 8.8 for a wall width of 10 cm and 20 cm. We can read on the figure that the corresponding transmission coefficients are approximately

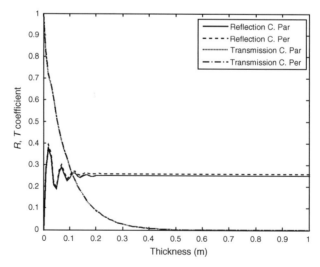

Figure 8.8 Generalized reflection and transmission coefficients as a function of the thickness for an incidence angle of $10°$ with $\varepsilon = 2.5\varepsilon_0$, $\sigma = 0.01$ and $f = 2$ GHz

0.25 and 0.08, which in dB (20log) mean $-12\,\text{dB}$ and $-22\,\text{dB}$, respectively. In this case, the incidence angle was $10°$, and $\varepsilon = 2.5\varepsilon_0$, $\sigma = 0.01$ and $f = 2\,\text{GHz}$.

The previous model can give a good approximation of the transmission loss when the wall structure and material properties are fairly homogeneous. However, it is quite common practice to use empirical propagation models that include empirically derived wall transmission loss values as discussed in the following section.

8.3 Microcell and Picocell Empirical Models

In several of the simulation projects proposed in the next section, we will be considering a deterministic approach to microcell and indoor picocell propagation modeling. We will do this by performing ray tracing and coherent addition of rays. To further improve the modeling, diffraction effects can be introduced. This will complement the theoretical presentation provided in the previous section, where basic propagation mechanics such as reflection and transmission are analyzed. For completeness, here we provide a brief overview of the empirical models used in these channels for assessing the average path loss.

In *microcells* [3], measurements have shown that two distance-dependent decay rates can be observed. Thus, the path loss can be described using the expression, $l(d) = kd^n$, similar to what we saw for macrocells, where n will show two distinct values. Typically, the constant k represents the loss at the reference distance (1 or 100 m, for example) and can be equated, without introducing significant errors, to the free-space loss. Thus, typical microcell models follow the general expression,

$$L(d) = L_{\text{fs}}(1\,\text{m}) + 10n \log d \tag{8.18}$$

where d is in meters, and L and L_{fs} in dB. To introduce the two slopes, a second term must be added for distances beyond the so-called *break point*, d_{B}, and hence,

$$
\begin{aligned}
l_1(d) &\alpha d^{n_1} && \text{for } d < d_{\text{B}} \\
l_2(d) &\alpha d_{\text{B}}^{n_1}\left(\frac{d}{d_{\text{B}}}\right)^{n_2} && \text{for } d > d_{\text{B}}
\end{aligned}
\tag{8.19}
$$

where α means proportional to. In order to introduce a smooth transition between the two slopes, an appropriate combination formula must be used to gently switch from l_1 to l_2,

$$l(d) = \{[l_1(d)]^p + [l_2(d)]^p\}^{1/p} \tag{8.20}$$

where parameter p is a so-called *shape factor* that determines how abrupt the transition between both slopes is. Typically, p is in the order of 4. Note that, at the break point, d_{B}, both loss values, l_1 and l_2, are equal.

Similarly, after turning a corner into a side street (Figure 8.9) MS goes into the NLOS (non-line-of-sight) state. The path loss, in this case, will be modeled in much the same way

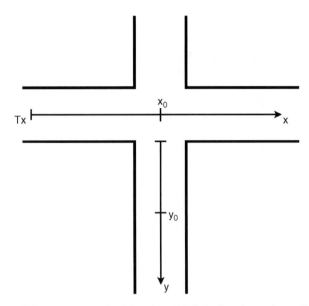

Figure 8.9   Geometry of LOS and NLOS links in urban microcells [3]

as in the LOS case by assuming a hypothetical BS sited at a position facing the NLOS street. The overall loss will be the sum (in dB) of the loss until the street junction and the loss within the side street. We will also be simulating this situation in the section on projects, using a deterministic approach where the reflection and diffraction mechanisms are combined.

Typical model parameter values for LOS paths at 1800 MHz are in the order of $n_1 \approx 2.3$, $n_2 \approx 5.6$ and $d_B \approx 250$ m. For NLOS paths, larger exponents have been reported for both decay rates.

Indoor propagation conditions [3] differ considerably from those found outdoors. The distances between the transmitter and the receiver are much smaller due to the high attenuations introduced by walls, floors and the furniture, and also to the smaller transmit powers used. Empirical models normally take into account different path types:

- LOS (line-of-sight) paths where BS and MS are in total or partial view of each other, e.g., in open rooms or halls with or without blocking objects.
- OLOS (obscured line-of-sight) paths where thin walls or objects block the direct signal.
- NLOS (non-line-of-sight) paths where thick walls block the direct signal.

Different expressions have been proposed for the average path loss. Some expressions are based on a distance power law, as in macrocells, i.e., $L(d) = 10n \log(d) + K_1$ (dB), where typical values of n are about 3 for LOS paths, 4 for OLOS and 6 for NLOS. Other models use linear attenuation coefficients, i.e., $L(d) = 20 \log(d) + \alpha d + K_2$ (dB), where the free-space loss, i.e., $n = 2$, is taken as the reference, and α is in dB/m. Typical values of α range from 0.2 to 0.6.

Other empirical models use more detailed input data, for example, the so-called COST 231 *multi-wall model* [4], which predicts the path loss as the sum of the free-space loss and the loss due to all walls and floors traversed. The expression for this model is

$$L = L_{fs} + L_C + \sum_{i=1}^{l} k_{w-i} L_{w-i} + k_f^{\left(\frac{k_f+2}{k_f+1} - b\right)} L_f \tag{8.21}$$

where L_{fs} is the free-space loss, L_C is a constant loss, k_{w-i} is the number of walls of type i traversed, k_f is the number of floors traversed, L_{w-i} is the type i wall loss, L_f is the floor loss, b is an empirical parameter and l is the number of wall types considered. Walls can be roughly classified into *light walls*, i.e., walls not bearing load, e.g. plasterboard or light concrete walls, and *heavy walls*, i.e., load-bearing walls, e.g., thick walls made of concrete or brick. Parameter values quoted in [4] for 1800 MHz are $L_W = 3.4$ for light walls and $L_W = 6.9$ for heavy walls, $L_f = 18.3$ dB and $b = 0.46$.

We now consider empirical models dealing with *outdoor-to-indoor* paths. For an adequate development of cellular networks, it is of great interest to provide a good coverage quality within buildings using BSs located outdoors.

To compute the total loss, the following expression can be used

$$L_T(dB) = L(d) + L_{we}(v_i, v_h) + n_w L_{wi} - n_f G_h \tag{8.22}$$

where L_T is the total path loss, $L(d)$ is the distance-dependent path loss up to the building, L_{we} is the external wall attenuation, L_{wi} is the internal wall attenuation, n_w is the number of internal walls, n_f is the floor number, where zero is the ground floor, G_h is the height gain per floor, typically equal to 2 dB/floor, v_i is the incidence angle, and v_h is the deviation from horizontal plane (Figure 8.10). Typical wall-loss values range from a few dB up to 20 dB.

8.4 Projects

In the following, we present a number of projects that try to provide the reader with a means of understanding some of the various aspects, specific to microcell and picocell propagation, that differ from the macrocell case, which has been extensively analyzed in previous chapters in the examples and simulations proposed.

Project 8.1: The Two-Ray Model

We perform simple ray tracing in this and some of the following projects. We start by reproducing the *plane-earth model*, which can be considered as a good approximation of microcell propagation conditions when side reflections are not significant. This also occurs, for example, in highway microcells. This model has also been frequently used to explain why the power decay rate in macrocells is closer to an $n = 4$ inverse distance law than to $n = 2$ (free space).

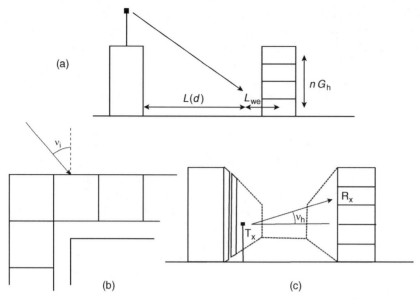

Figure 8.10 (a) Model geometry. (b) Top view. Incidence angle. (c) Side view. Horizontal plane deviation [3]

The idea in this project is tracing and computing the direct and reflected fields, and adding them up coherently. The direct field is given by

$$e_0 = \sqrt{60\,p\,g/l}\ \frac{\exp(-\mathrm{j}k\,d_0)}{d_0} \tag{8.23}$$

and the reflected signal by

$$e_\mathrm{R} = \sqrt{60\,p\,g/l}\,R\,\frac{\exp[-\mathrm{j}k(d_1 + d_2)]}{d_1 + d_2} \tag{8.24}$$

where p is the power, g is the antenna gain and l the loss at the transmit side – all in linear units. Parameter R is the reflection coefficient. To make things easier, we are going to assume that we are operating near gazing incidence, and thus R is approximately -1 regardless of the polarization for an average ground and for all simulated positions of the receiver (MS route). This is a fairly good approximation anyway, as can be verified using the reflection coefficient scripts provided (`intro81` and `intro82`). Note that the field strengths are peak values. To produce rms values, replace $\sqrt{60}$ with $\sqrt{30}$.

We have implemented a simulator (`project811`) where the receiver travels away from the transmitter. The settings used were as follows: the frequency was 2 GHz, the transmit power 1 Watt, the transmit gain 0 dBi (ratio equal to one), the BS antenna height was 10 m and the MS antenna height was 1.5 m. The sampling spacing was 1 m. Very simple ray tracing has been implemented, basically involving the calculation of the distance from the

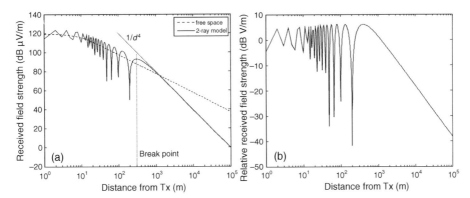

Figure 8.11 (a) Received field strength as a function of the distance from BS in a microcell. (b) Normalized field strength with respect to the direct signal as a function of the distance from BS

transmit and the receive antennas (d_0) and the distance from the ground image of the transmit antenna ($-h_t$) to the receive antenna ($d_1 + d_2$).

Figure 8.11(a) shows the overall field strength in dBμV/m (the usual practical unit), i.e., $E(\mathrm{dB}\mu\mathrm{V}/\mathrm{m}) = 20\log(e \times 10^6)$, where e is the calculated total field strength in V/m. Figure 8.11(a) also shows the evolution of the direct field for reference. We see that, close to the transmitter, strong oscillations about the direct signal level take place. After a given distance, the so-called *break point*, the decay rate is more pronounced than in free space, and corresponds to an $n = 4$ law. The location of the break point and the reason for the $n = 4$ law are discussed later in some detail.

The above means that the received signal decays very fast after the break point. This is very good interference-wise, since a microcell will be isolated from nearby *cochannel* cells after a relatively short distance. Figure 8.11(b) shows the field strength relative to the direct signal.

We have also simulated a vertical antenna height scan at the MS side in `project812`. The settings were the same as in the previous project, except for the BS–MS distance, now constant and equal to 200 m. The elevation step was 0.1 m. Figure 8.12 illustrates the field strength in practical units vs. the MS antenna height. The free-space level is also plotted for reference. Again, a pattern with several lobes is observed, corresponding to the interference between the direct and the reflected rays. Below a given height, the overall signal seems to take up values well below that of the direct ray, as in the previous case for the longer distances.

Figure 8.13 illustrates the geometry of the two-ray model also called the *plane-earth model*. This yields an $n = 4$ inverse power path loss propagation law, which is very similar to that observed in macrocells (Chapters 1 and 3). The total received field strength is

$$e = \sqrt{60pg/l}\left\{ \frac{\exp(-\mathrm{j}k\,d_0)}{d_0} + R\frac{\exp[-\mathrm{j}k(d_1 + d_2)]}{d_1 + d_2} \right\}$$
$$= e_0[1 + R\exp(-\mathrm{j}\Delta\phi)] \tag{8.25}$$

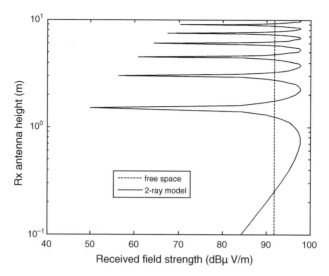

Figure 8.12 Received field strength as a function of the MS antenna height for a constant distance from BS in a microcell

where R is the ground reflection coefficient, $R = |R| \exp(-\mathrm{j}\beta)$, and e_0 is the direct ray field strength. The phase difference due to different paths lengths is $\Delta\phi = 2\pi\Delta l/\lambda$, where Δl is the path difference. Its value is

$$
\Delta l = \mathrm{TxPRx} - \mathrm{TxRx} = \left[d^2 + (h_\mathrm{t} + h_\mathrm{r})^2\right]^{1/2} - \left[d^2 + (h_\mathrm{t} - h_\mathrm{r})^2\right]^{1/2}
$$
$$
= d\left[1 + \left(\frac{h_\mathrm{t} + h_\mathrm{r}}{d}\right)^2\right] - d\left[1 + \left(\frac{h_\mathrm{t} - h_\mathrm{r}}{d}\right)^2\right] = \frac{2h_\mathrm{t}h_\mathrm{r}}{d} \tag{8.26}
$$

Using the first two terms of the series expansion

$$
(1 + x)^\alpha \approx 1 + \frac{\alpha}{1!}x + \frac{\alpha(\alpha - 1)}{2!}x^2 + \cdots \tag{8.27}
$$

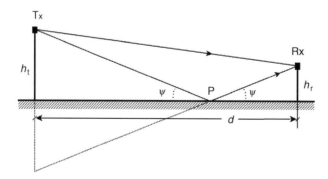

Figure 8.13 Plane-earth model

particularized for the square root, i.e., $\alpha = 1/2$, we get

$$\Delta l \approx d \left[1 + \frac{1}{2} \left(\frac{h_\mathrm{t} + h_\mathrm{r}}{d} \right)^2 \right]^{1/2} - d \left[1 + \frac{1}{2} \left(\frac{h_\mathrm{t} - h_\mathrm{r}}{d} \right)^2 \right]^{1/2} \tag{8.28}$$

Coming back to the expression for the total received field strength and introducing the magnitude and phase of the reflection coefficient, the total field strength is

$$e = e_0 \{ 1 + |R| \exp[-j(\Delta\phi + \beta)] \} \tag{8.29}$$

and its magnitude is

$$|e| = |e_0| \left[1 + |R|^2 + 2|R| \cos(\Delta\phi + \beta) \right]^{1/2} \tag{8.30}$$

Taking into account that

$$\sin^2(\alpha) = \frac{1 - \cos(2\alpha)}{2} \tag{8.31}$$

and that $d \gg h_\mathrm{t}, h_\mathrm{r}$, i.e., $\psi \approx 0$, with $|R| \approx 1$ and $\beta \approx \pi$, the magnitude of the total received field is

$$|e| = 2|e_0| \left| \sin\left(\frac{\Delta\phi}{2} \right) \right| \approx 2|e_0| \left| \sin\left(\frac{2\pi h_t h_r}{\lambda d} \right) \right| \tag{8.32}$$

From this expression, it is clear that the combination of a direct and a reflected ray gives rise to an interference pattern with amplitudes ranging from 0 to $2e_0$. Figure 8.12 illustrates the received field relative to the direct signal when an antenna height scan is carried out on the receive side. In both macrocell and microcell environments, low antennas are used, i.e., $h_\mathrm{t}, h_\mathrm{r} \ll d$, the argument within the sine function is small, and thus, the sine can be replaced by its argument, yielding

$$|e| \approx |e_0| \frac{4\pi h_t h_r}{\lambda d} \tag{8.33}$$

In other words, when the above conditions are fulfilled, the receiver will be moving below the first maximum of the interference pattern shown in Figures 8.11 and 8.12. The excess loss, in this case, is

$$l_{\mathrm{excess}} = \frac{|e_0|^2}{|e|} \approx \left(\frac{\lambda d}{4\pi h_t h_r} \right)^2 \tag{8.34}$$

The overall path loss, expressed in linear units, is the product of the free space and the excess loss, i.e.,

$$l = l_{\text{fs}}\, l_{\text{excess}} = \left(\frac{4\pi d}{\lambda}\right)^2 \left(\frac{\lambda d}{4\pi h_t h_r}\right)^2 = \left(\frac{d^2}{h_t h_r}\right)^2 \tag{8.35}$$

Note that the wavelength (or the frequency) has disappeared, i.e., there is no frequency dependence in the plane-earth loss, unlike what happens in the free-space case. It can also be observed how the path-loss dependence with the inverse of the distance follows an $n = 4$ power law, that is, a 40 dB/decade decay rate. In practical units, the model expression is

$$L_{\text{pe}}(\text{dB}) = 120 + 40 \log d(\text{km}) - 20 \log[h_t(\text{m}) h_r(\text{m})] \tag{8.36}$$

The change in behavior from $n = 2$ to $n = 4$ is usually identified with the last maximum in the overall field. This coincides with an argument of the sine function equal to $\pi/2$. The point of occurrence of the last maximum, as said, is called the *break point* and is located at a distance

$$d_{\text{B}} = \frac{4 h_t h_r}{\lambda} \tag{8.37}$$

Project 8.2: Street Canyon Propagation

In `project82`, we have implemented a more realistic simulator for an urban microcell. This model can also be called the *urban canyon* or *street canyon* model. In this case, we consider the existence of reflections on the building faces at both sides of the street. Street canyon propagation can, hence, be characterized by the so-called *four-ray model*: direct ray, ground reflection, wall-1 and wall-2 reflections, or the *six-ray model* where double building wall reflections are also taken into account, as illustrated in Figure 8.14. Other, more complex models involving double ground-wall reflections can also be envisaged.

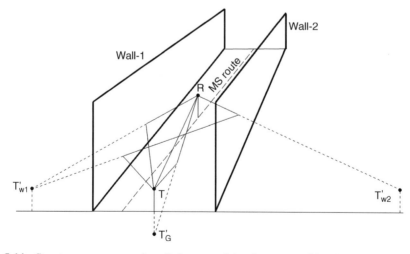

Figure 8.14 Street canyon propagation. Only some of the six rays considered are shown to make the figure clearer

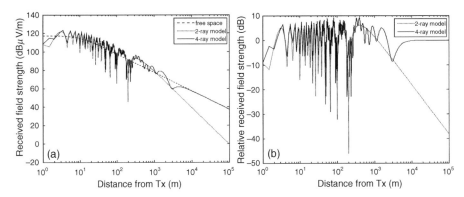

Figure 8.15 (a) Absolute field strengths for the two- and four-ray models. (b) Normalized field strengths for the two- and four-ray models

In our implementation, we have assumed a ground reflection coefficient of -1 and a wall reflection coefficient of -0.5, taking into consideration that a vertical polarization is transmitted. The reader is encouraged to check these values using the scripts provided (`intro81` and `intro82`). As for the ray tracing, it is again carried out using *image theory* as also illustrated in Figure 8.14.

Figure 8.15(a) shows the resulting absolute field strengths for the two- and four-ray models while in Figure 8.15(b) we show the corresponding normalized fields. The same type of plots is shown in Figure 8.16 for the four- and six-ray models. Additionally, Figure 8.17 illustrates the resulting ideal (deltas) power delay profile for the six-ray model at a distance of 100 m. The resulting PDP parameters for the four- and six-ray models: average delay (D) and delay spread (S) were obtained using function `PDPparameters`. The results obtained are as follows:

```
4-rays model
Excess delay   6.59519e-009
Delay spread   3.90613e-009
```

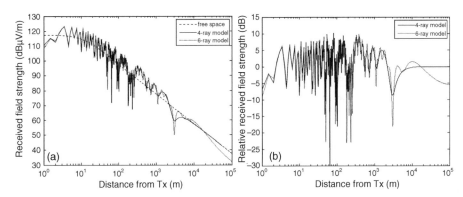

Figure 8.16 (a) Absolute field strengths for the four- and six-ray models. (b) Relative field strengths for the four- and six-ray models

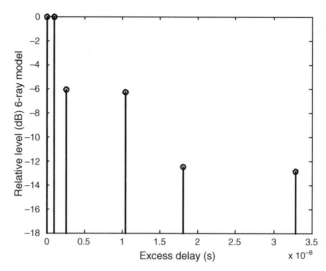

Figure 8.17 Resulting ideal (deltas) power delay profile for the six-ray model at a distance of 100 m

```
6-rays model
Excess delay  1.9342e-008
Delay spread  1.10096e-008s
```

Project 8.3: Side Street Scenario

In this project we try and simulate the case where MS enters a side street. We have simulated the simplified urban geometry depicted in Figure 8.18. In this case we will introduce, in addition to the reflected rays, diffraction contributions where we will perform the integration of the Fresnel sine and cosine functions throughout the side street opening, as in Chapter 2. To simplify things, we will perform the integration from the ground level to infinity along the z-axis, while along the y-axis the integration will be between wall-3 and wall-4. Possibly some of the simplifying assumptions made in Chapter 2 are not totally fulfilled. However, this approach will provide us with a rough estimate of the diffraction loss. As shown in the figure, we have considered, for the main street along the y-axis a four-ray model, except for the missing rays on wall-2 at the side street opening. Along the side street on the x-axis, we have performed the complex addition of several diffracted (through the side street opening) contributions involving the transmitter, T, its images with respect to the ground, T'$_G$, and walls-1 and -2, the receiver, R, and its images with respect to the ground, R'$_G$, and walls-3 and -4.

Figure 8.19 shows the actual geometry simulated and Figure 8.20(a) shows the overall path gain (inverse of the path loss) for both the main and side street MS routes. Finally Figure 8.20(b) shows the relative signal levels with respect to free space. From these figures, it is clear how turning into a side street brings about very pronounced attenuation levels. This fact was already mentioned in Section 8.3 when we discussed empirical microcell models.

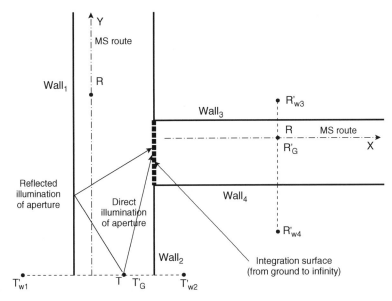

Figure 8.18   Project 8.3 geometry

Project 8.4: Wideband Indoor Propagation – The Saleh–Valenzuela Model

Here we provide an example of a statistical wideband channel model for indoor picocells. The *Saleh–Valenzuela model* [5], S&V, is a very well-accepted model for this propagation scenario. It has also been extended to include the directional properties of the channel for use in *multiple-input-multiple-output* (MIMO) applications, as we will see in Chapter 10. We

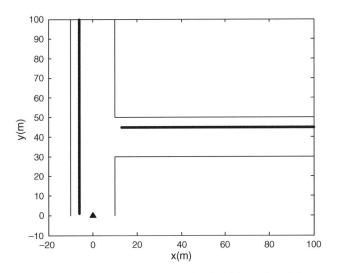

Figure 8.19   Actual geometry simulated in Project 8.3

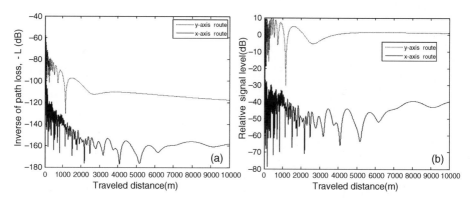

Figure 8.20 (a) Path gain for both the main and side street MS routes. (b) Relative signal levels with respect to free space for both the main and side street MS routes

implemented in `project84` this model which we describe below. It is hoped that this discussion, and the use and editing of our implementation, will provide the reader with a good insight into the indoor channel. Maybe the most interesting concept introduced is that of ray or echo *clustering*, that is, multipath contributions tend to reach the receiver in groups with similar delays, and angles of departure and arrival. These groups can very well correspond to contributions from the same element in the propagation environment such as a piece of furniture, for example.

As said, the S&V model makes the assumption that *echoes* arrive in *clusters*, the overall impulse response being

$$h(\tau) = \sum_l \sum_k \beta_{kl} \exp(j\phi_{kl})\delta(\tau - T_l - \tau_{kl}) \tag{8.38}$$

where index l indicates the *cluster number* and index k indicates the *echo number within a cluster*. The amplitude (representing the received voltage) of each echo is β_{kl} (Figure 8.21).

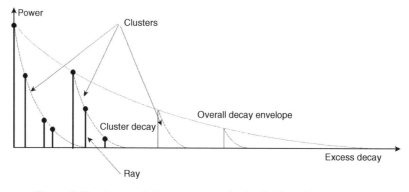

Figure 8.21 Assumed cluster structure in the Saleh–Valenzuela model

The received echo structure presents a *decay rate* defined by $\overline{\beta_{kl}^2}$, a *negative exponential law* for powers expressed in linear units or, alternatively, a linear law for powers expressed in dB,

$$\overline{\beta_{kl}^2} = \overline{\beta^2(0,0)} \exp(-\mathrm{T}_l/\Gamma) \exp(-\tau_{kl}/\gamma) \tag{8.39}$$

where

- $\overline{\beta^2(0,0)}$ is the average power of the first arrival in the first cluster (linear units);
- γ is the decay rate within each cluster;
- Γ is the cluster decay rate (envelope) (Figure 8.21);
- τ_{lk} are the ray times of arrival within cluster l; and
- T_l are the cluster times of arrival.

The number of arrivals and times of arrival for both clusters and echoes are characterized by means of *Poisson* and *exponential* distributions. The *cluster arrival rate*, Λ, denotes the parameter for the *inter-cluster arrival times*, and the *echo arrival rate*, λ, refers to the parameter for the *intra-cluster arrival times*. The distributions of these arrival times are as follows,

$$p(\mathrm{T}_l|\mathrm{T}_{l-1}) = \Lambda \exp[-\Lambda(\mathrm{T}_l - \mathrm{T}_{l-1})] \quad \mathrm{T}_{l-1} < \mathrm{T}_l < \infty \tag{8.40}$$

and

$$p(\tau_{kl}|\tau_{(k-1)l}) = \lambda \exp[-\lambda(\tau_{kl} - \tau_{(k-1)l})] \quad \tau_{(k-1)l} < \tau_{kl} < \infty \tag{8.41}$$

These two distributions are assumed to be independent of each other. Note that the arrivals are conditioned or referred to the previous arrival. The first cluster arrival can be set to zero. This would mean that the rest of the arrivals represent excess delays with respect to the first arrival.

Next, we discuss the simulation procedure implemented in `project84`. The received power, assuming a receive antenna gain of 0 dBi, was calculated considering a power decay law $n = 4$. We calculated the free-space loss at 1 m from the transmitter, and propagated this power onward assuming exponent n, i.e.,

$$\overline{P_{\mathrm{r}}}(d) = \mathrm{EIRP} - L_{\mathrm{fs}}(1\mathrm{m}) - 10n \log(d) \, \mathrm{dBm} \tag{8.42}$$

At the distance of interest, d, we normalized the received power with respect to that for free space, i.e., $\bar{p}'(d) = \bar{p}_{\mathrm{r}}(d)/p_{\mathrm{fs}}(d)$ where $\bar{p}'(d)$ is the normalized average power and $\bar{p}_{\mathrm{r}}(d)$ is the actual average received power predicted by the propagation model.

Then, we generated the cluster structure by drawing a sufficiently large number of arrivals: 100 in this case. Not all clusters will contribute a significant amount of power, and those will be dropped. To do this, after generating the arrivals using the exponential inter-arrival distribution shown above, we attribute to all these arrivals a normalized power equal to one (Figure 8.22(a)). Then, we applied the exponential decay law (Figure 8.22(b)) and computed the overall normalized power. This will not match the wanted value, $\bar{p}'(d)$. Thus, the

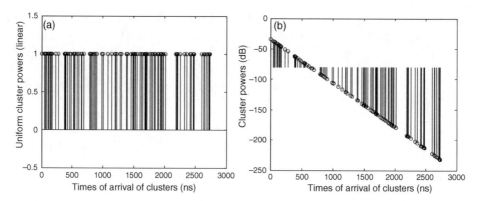

Figure 8.22 (a) Step where cluster arrivals are generated and assigned the same power. (b) Cluster arrivals plus cluster decay rate

difference in dB was calculated, and the whole structure was corrected for by shifting it up or down by the power difference in dB. Then, only the first clusters were kept, the criterion being that the overall normalized power was within 0.5 dB of the required objective, $\bar{p}'(d)$ (Figure 8.23). This allows disregarding most of the nonsignificant clusters. The process is repeated within each individual cluster. Figure 8.24(a) shows the generated cluster structure and Figure 8.24(b) shows the overall ray structure within the clusters generated in the previous step.

The exponential random number generator used (genExponential) was based on the relationship between the Rayleigh and the exponential distribution that we saw in Chapter 5. Thus, we generated a Rayleigh series and squared the resulting series, taking into account

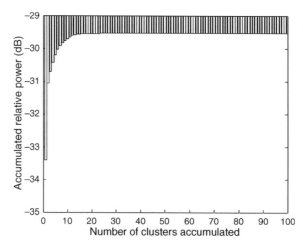

Figure 8.23 Accumulation of cluster powers. We drop those clusters with negligible contributed power. The overall being within 0.5 dB of the objective, in this case -29.5 dB

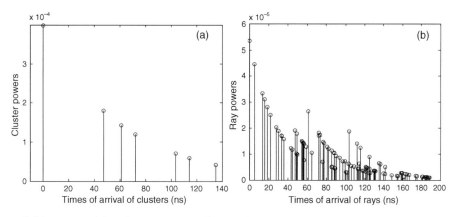

Figure 8.24   (a) Resulting cluster structure with associated powers amounting to the total normalized received average power. (b) Resulting structure with individual echoes within previously computed clusters

that the rms squared value of the Rayleigh distribution is equal to the mean (and also the standard deviation) of the exponential.

Finally, time series have been produced for each ray. For this, we have assumed as indicated by the S&V model, a Rayleigh distribution. We have equated the relative average power for ray k within cluster l to the rms squared value of the Rayleigh distribution, $2\sigma^2$ (Chapter 1). The following power budget is fulfilled,

$$\bar{p}'(d) = \sum_l \sum_k \bar{p}'_{kl} \tag{8.43}$$

where $\bar{p}'_{kl}$ is the corresponding normalized average ray power obtained through the process outlined above.

To generate all the ray series, we have assumed very slow time variations, e.g. due to people moving by, characterized by a Butterworth Doppler filter (as we did in Project 5.4) with the following settings: sampling frequency 20 Hz, `Wp=0.01`, `Ws=0.2`, `Rp=3` and `Rs=40`, where `Wp` is the normalized frequency with respect to `fs/2` indicating the end of the pass-band, `Ws` is the normalized frequency with respect to `fs/2` indicating the beginning of the stop-band, `Rp` is the pass-band attenuation in dB and `Rs` is the stop-band attenuation in dB. Other settings were $\gamma = 29\,\text{ns}$, $\Gamma = 60\,\text{ns}$, and $1/\lambda = 5\,\text{ns}$ and $1/\Lambda = 300\,\text{ns}$. The simulation of each of the series was carried out using `genRayleighFiltered` which follows the same procedure as in Project 5.4. Figure 8.25 shows all the ray time series generated.

Project 8.5: Building Penetration Through Windows

In this project, we try to simulate the propagation from the outside to the inside of a building through a window, which we model as an aperture. To do this, we come back to the techniques discussed in Chapter 2 regarding diffraction effects. There, we were more

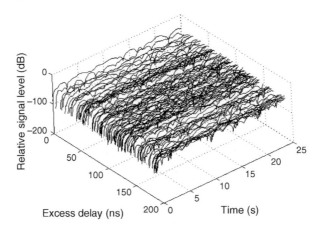

Figure 8.25   All ray time series corresponding to the multipath structure previously generated

concerned with semi-infinite apertures. Now, we are interested in bounded apertures as illustrated in Figure 8.26. We remind the reader of the formula (Equation 2.32) for calculating the diffraction loss due to a plane, rectangular aperture, i.e.,

$$F_d(u,v) = F_d(u)F_d(v) = \sqrt{\frac{j}{2}}\{[C(u_2) - C(u_1)] - j[S(u_2) - S(u_1)]\}$$

$$\times \sqrt{\frac{j}{2}}\{[C(v_2) - C(v_1)] - j[S(v_2) - S(v_1)]\} \qquad (8.44)$$

The settings for `project85` were as follows: the frequency was 2 GHz, the transmitter position outside the building was `xt=100`, `yt=50` and `zt=50`. The receiver was located

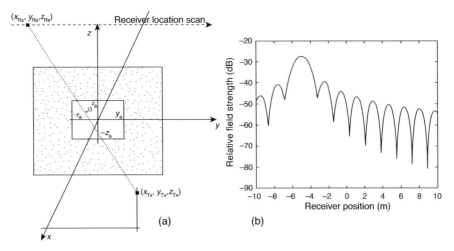

Figure 8.26   (a) Simulated geometry. (b) Relative field strength propagating through a window

at `xr=-10`, `zr=0`, while the `yr` position changed from -10 to 10 m with a step of $\lambda/8$. The window is defined by `ya1=-0.5`, `ya2=0.5`, `za1=-0.4`, `za2=0.4`. In Figure 8.26(a) point O indicates the intersection of the direct path with the plane containing the window. Some simple geometry can be used for calculating the location of O (y_0,z_0). Thus, the direct ray between transmitter and receiver is described by the straight-line equation

$$\frac{x - x_t}{x_r - x_t} = \frac{y - y_t}{y_r - y_t} = \frac{z - z_t}{z_r - z_t} \tag{8.45}$$

which, particularized at the plane containing the window, i.e., $x = 0$, becomes

$$\frac{-x_t}{x_r - x_t} = \frac{y_0 - y_t}{y_r - y_t} = \frac{z_0 - z_t}{z_r - z_t} \tag{8.46}$$

From this equation, we can calculate y_0 and z_0, i.e.,

$$y_0 = \frac{-x_t(y_r - y_t)}{x_r - x_t} + y_t \text{ and } z_0 = \frac{-x_t(z_r - z_t)}{x_r - x_t} + z_t \tag{8.47}$$

Finally, Figure 8.26(b) shows the resulting relative field strength, where a maximum is found right in front of the transmitter and, as the receiver moves, we can observe marked lobes. The reader is encouraged to change the frequency and the size of the window to observe the effects of these changes. Note, however, that this is assuming a window on a perfectly absorbing screen. One physical model that could help us quantify the loss through the wall would be the slab model discussed in Section 8.2.

8.5 Summary

In this chapter, we have reviewed several issues of relevance for the microcell and picocell scenarios. We have also considered the outdoor-to-indoor propagation case. In addition to reviewing some basic theoretical and empirical techniques, we have proposed and implemented simulations which deal with the modeling of these scenarios using simple, *image theory ray tracing* techniques which reproduce what empirical models forecast. Moreover, we have introduced diffraction effects in these scenarios. Finally, we have also presented a widely used statistical model, due to Saleh and Valenzuela, for describing the wideband indoor channel, where we have introduced the concept of ray clustering.

References

[1] ITU-Recommendation P.527-3. Electrical characteristics of the surface of the Earth, 1992.
[2] H.L. Bertoni. *Radio Propagation for Modern Wireless Systems*. Prentice Hall, 2000.
[3] J.M. Hernando & F. Pérez-Fontán. *An Introduction to Mobile Communications Engineering*, Artech House, Boston, 1999.
[4] Evolution of land mobile radio (including personal) communications. Euro-COST 231 Project. Final Report, 1996.
[5] A.A.M. Saleh & R.A. Valenzuela. A statistical model for indoor multipath propagation. *IEEE J. Select. Areas Comm.*, **SAC-5**(2), 1987, 128–137.

Software Supplied

In this section we provide a list of functions and scripts developed in MATLAB®
(MATLAB® is a registered trademark of The MathWorks, Inc.) implementing the various
projects and theoretical introductions mentioned in this chapter. They are the following:

```
intro81              Fresnel_integrals
intro82              rfresnel
intro83              tfresnel_g
project811           PDPparameters
project812           genExponential
project82            cprop
project83            genRayleighFiltered
project84            s_d
project85            ep
                     stem2D
                     erfz
                     rayleigh
                     tfresnel
                     filtersignal
                     rfresnel_g
```

9

The Land Mobile Satellite Channel

9.1 Introduction

Coverage of extended service areas (*megacells*), especially in remote places, can only be achieved by using satellites either in *geostationary orbit* (GEO) or in non-GEO orbits: *low earth orbit* (LEO), *medium earth orbit* (MEO), *highly elliptical orbit* (HEO), etc. This is the case of radio and TV broadcast systems to users on the move, or cellular satellite systems. Also, satellite navigation systems such as the *global positioning system* (GPS) are affected by similar propagation conditions. Typically L- (1–2 GHz) and S-bands (2–4 GHz) are used for *land mobile satellite* (LMS) services.

LMS propagation is affected in different ways by the *ionosphere*, the *troposphere* and the *environment* surrounding the mobile receiver. We address here the modeling of *environmental effects*, that is *shadowing* and *multipath*. At these frequencies, the troposphere causes very small attenuation due to gasses or rain, and their effects can be assumed to be negligible, while the ionosphere gives rise to *Faraday rotation*, which is still noticeable at these bands. This is one of the reasons why circular polarization is used at these frequencies. Polarization rotations of up to 48° can be observed at L-band in worst-case scenarios. This rotation would cause unacceptable polarization mismatches for linearly polarized antennas. For mobile terminals, both vehicular or handheld, non-directive antennas will normally be used, showing hemispherical patterns with very little directivity.

An important element in LMS systems is that a single satellite is not sufficient for achieving the wanted coverage reliability but rather constellations of several satellites must be employed. This allows the use of *satellite diversity* for improving system availability. If a link with one of the satellites is interrupted by shadowing, an alternative satellite should be present to help reduce the outage probability.

To model the effects of the relative changing positions of the satellites in non-GEO constellations, *orbit simulators* can be used. These programs include as inputs the current time and location of the terminal together with the appropriate *Kepler parameters* of the satellites in the constellation. Together with the MATLAB® (MATLAB® is a registered trademark of The MathWorks, Inc.)

Modeling the Wireless Propagation Channel   F. Pérez Fontán and P. Mariño Espiñeira
© 2008 John Wiley & Sons, Ltd

simulators mentioned in this book, we provide an orbit simulator, `generator`, running in MATLAB®, which provides satellite trajectories. The results are stored in. MAT files. Kepler parameters (elements) for existing satellite constellations are periodically updated and can be downloaded from the WEB, for example at http://celestrak.com/NORAD/elements/. The file format normally used is the so-called NASA two-line element (tle). There are many web pages which can be browsed to learn more on the topic of satellite orbits.

The overall received signal is made up of the *direct signal* and *multipath*. In general, the time dispersion caused by the LMS channel will be fairly moderate, this is why here we concentrate on its narrowband characteristics. The *direct signal* may reach the receiver under line-of-sight (LOS) conditions, via diffraction on buildings, or absorption/scattering through vegetation. *Diffraction* contributions are mainly generated on the horizontal edges of the buildings in the street MS is traveling through. *Absorption/scattering* through trees greatly depends on the path length through the canopies, the density of foliage and branches. The characterization of tree attenuation is normally carried out by means of a tree-type-dependent *specific attenuation*, γ (dB/m). Typical measured γ values at L-band are in the range of 0.7 to 2 dB/m.

Multipath can be made up of a specular and a diffuse component. The *specular component* is caused by reflections on smooth surfaces on natural or man-made obstacles, or the ground. Polarization inversions may occur if the *Brewster angle* (Chapter 8) is exceeded, thus the specular reflected rays being cancelled out by the cross-polar radiation pattern of the receive antenna.

Diffuse scattered multipath may be generated on a number of environmental features. The irregular terrain gives rise to non-specular reflections, also building faces and trees may be the sources of this type of multipath. The distance decay of diffuse multipath is much faster than that of specular reflections, as seen in Chapter 7. The area around the receiver generating this kind of multipath will be limited to a few hundreds of meters, except for extremely large structures with very high *radar cross-sections* (RCS).

Other elements in the environment giving rise to shadowing and/or multipath effects can be overpasses, bridges and other features near MS as, for example utility poles, traffic lights, traffic signs, parked or moving vehicles, etc.

9.2 Projects

In this section we try to simulate some important features of the LMS channel.

Project 9.1: Two-State Markov Model

A suitable model for the narrowband LMS channel is a state-oriented approach, given that single distributions, e.g., Rice, Rayleigh, Loo [1], cannot describe the large dynamic range observed in the received signal. Measured pdfs tend to show at least two modal values (two humps), indicating that they are best described as a combination of at least two distributions. Defining different states allows attributing different distributions to sections of the mobile route with different degrees of shadowing. The overall pdf in this case is given by

$$f_{\text{overall}}(r) = p_{\text{GOOD}} f_{\text{GOOD}}(r) + p_{\text{BAD}} f_{\text{BAD}}(r) \tag{9.1}$$

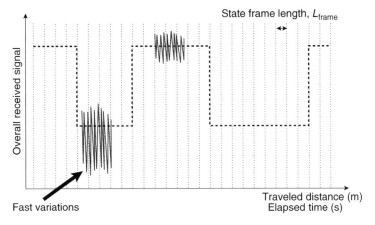

Figure 9.1   Two-state Markov model for the LMS channel [2]

Here, a two-state model [2], GOOD/BAD, is presented and simulated, where p_{GOOD} is the probability of being in the GOOD state and p_{BAD} is the probability of being in the BAD state, and f_{GOOD} and f_{BAD} are the pdfs describing the signal variations within these states.

There also exist many models postulating three or more states [3][4]. We will concentrate on a two-state one for simplicity's sake. The two possible states assumed here are:

- the *GOOD state* (line-of-sight and moderate shadow); and
- the *BAD state* (moderate and deep shadow).

Within each state, smaller scale variations are possible as illustrated in Figure 9.1. This model is narrowband since the time-spreading effects are not taken into consideration.

This type of model makes the following assumptions:

- The signal variations within each state are best described by several possible distributions, for example Rayleigh and Rice. More complex distributions like the ones discussed in Chapter 6, such as those of Suzuki [5] and Loo [1], or Corazza–Vatalaro [6], are frequently used.
- The transitions between states are modeled as a *first-order, discrete-time Markov chain*.

The parameters for each state, and the state and transition matrices for different environment types and elevations are calculated from measurements. *Absolute* state and state *transitions probability matrices*, **W** and **P**, model *state occurrences* and *durations*. The elements of the *state probability matrix*, p_i, fulfill the equation,

$$\sum_{i=1}^{2} p_i = 1 = p_{GOOD} + p_{BAD} \tag{9.2}$$

and the elements of the transition probability matrix, $p_{i|j}$, fulfill

$$\sum_{j=1}^{2} p_{i|j} = 1 \quad \text{with } i = 1 \text{ and } 2 \tag{9.3}$$

where $p_{i|j}$ is the probability of transition from state j to state i. The asymptotic behavior (convergence property) of the Markov chain is defined by the equation

$$\mathbf{P}\,\mathbf{W} = \mathbf{W}, \text{i.e.,} \begin{pmatrix} p_{1|1} & p_{1|2} \\ p_{2|1} & p_{2|2} \end{pmatrix} \begin{pmatrix} p_1 \\ p_2 \end{pmatrix} = \begin{pmatrix} p_1 \\ p_2 \end{pmatrix} \tag{9.4}$$

We also define a *minimum state duration* or *frame* (discrete-time Markov chain), L_{Frame}, of 1 m. Again, the term '*duration*' designates both *time durations* and *lengths of traveled route*. These two durations are linked by the *terminal speed*, V, which is assumed constant.

The durations of the states depend on the transition probabilities, $p_{i|j}$. Thus, the probability that the Markov chain stays in a given state, i, for n consecutive frames, or equivalently for $n \times L_{\text{frame}}$ meters (or $n \times L_{\text{frame}}/V$ seconds) is [7]

$$p_i(N = n) = p_{i|i}^{n-1}(1 - p_{i|i}) \quad \text{with } n = 1, 2, \ldots \tag{9.5}$$

and the cumulative distribution for the duration of each state is

$$p_i(N \leq n) = (1 - p_{i|i}) \sum_{j=1}^{n} p_{i|i}^{j-1} \quad \text{with } n = 1, 2, \ldots \tag{9.6}$$

An update of the current state is made each L_{frame} meters by drawing a random number.

The variations within the GOOD state are modeled by means of a Rice distribution, while for the BAD state a Rayleigh distribution is used. The autocorrelation of these fast variations is introduced by means of a Butterworth filter as we did in Project 5.4. A block diagram of the time-series generator based on this approach is shown in Figure 9.2 following the Lutz model [2].

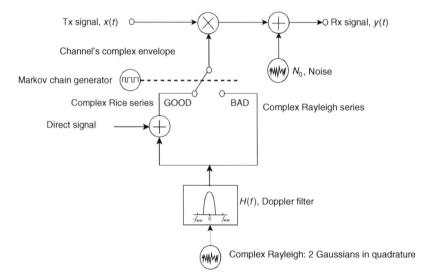

Figure 9.2   Two-state Markov plus Rayleigh/Rice LMS channel simulator [2]

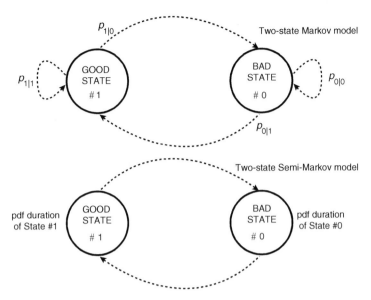

Figure 9.3　Markov vs. semi-Markov models

Other options exist for modeling state transitions and durations. A more flexible option than a *first-order Markov model* is using a *semi-Markov* [7] model (Figure 9.3). Typically, in this case, durations are modeled by a lognormal distribution.

Simulator `project91` has been run with the following settings for generating Figures 9.4–9.6: the carrier frequency was 1540 MHz, the frame length 1 m, the sampling fraction of the wavelength was 6. For the GOOD state, a standard deviation for the two Gaussian generators in quadrature was set to 0.2 (-11 dB/LOS $10\log(2\sigma^2)$), and for the

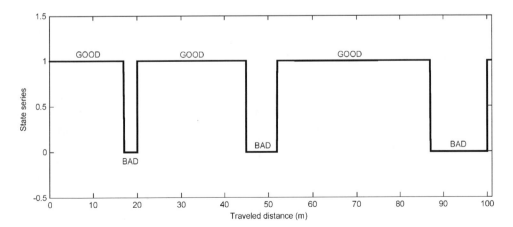

Figure 9.4　Good–Bad state series

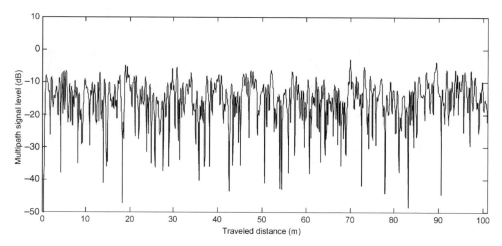

Figure 9.5 Multipath series for the state sequence in Figure 9.4

BAD state was set to 0.15 (-13.47 dB/LOS). A Butterworth filter with settings `Wp=0.09`, `Ws=0.16`, `Rp=3`, `Rs=50` was used for Doppler shaping. The transition matrix was

$$P = [0.95\ 0.05$$
$$0.1\ 0.9]$$

Figure 9.4 illustrates a series of states (1, GOOD, 0, BAD) drawn using the Markov model. Figure 9.5 shows the series of diffuse multipath where the two different σ values can be observed following the same evolution as the states. Note how the differences in the values of σ are barely perceptible. It is only when the direct signal is coherently added that the differences between the states are significant as shown in Figure 9.6. This figure shows the alternating Rice–Rayleigh series where the direct signal is included or not depending on the state.

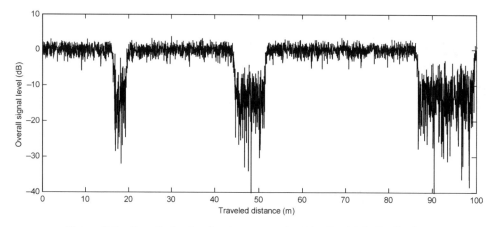

Figure 9.6 Overall simulated series: states plus Rice/Rayleigh distributions

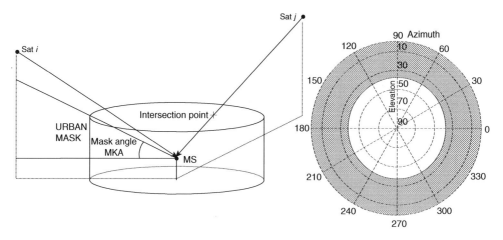

Figure 9.7   Cylindrical mask

Project 9.2: Coverage Using Constellations of Satellites

In this project, we want to get acquainted with the *satellite diversity* concept. We assume a static user terminal in an urban area, and identify the satellites in view at each point in time. If at least one satellite is visible, then the link will be in the *available state*.

The influence of the urban environment can be taken into account by means of a so-called *urban mask*. The simplest mask is one of constant *masking angle*, MKA. This is equivalent to considering that MS is in the center of a cylinder as illustrated in Figure 9.7. The same figure illustrates the mask in polar coordinates, azimuth and elevation, with a constant MKA of 40°.

Another common geometry used in many satellite diversity studies is the so-called *street canyon*, illustrated in Figure 9.8 with reference to the cylinder mask. A street canyon is a more realistic representation of the urban environment. In the same figure, we show its

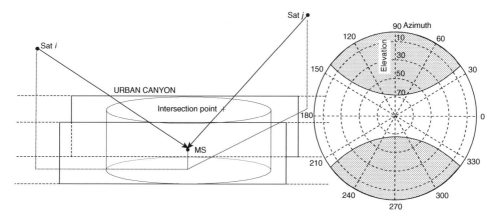

Figure 9.8   Street canyon and associated mask

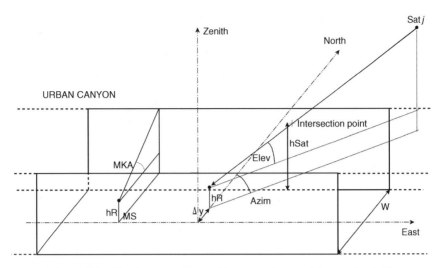

Figure 9.9 Street canyon. Geometric parameters used in `project92`

corresponding mask in polar coordinates. This mask concept draws from photogrammetric studies using zenith pointing fish-eye lenses [8].

We simulated in `project92` the canyon case. The calculations were made with reference to the geometric parameters shown in Figure 9.9. We use the constellation time series contained in `newSatMatrix.mat`. The reader is encouraged to study other constellations by running the constellation simulator `generator` plus script `picksats` which creates the file `newSatMatrix.mat` that stores, in this case of the file provided, the time evolution of 10 satellites. Each row corresponds to a point in time, the time spacing being of one second. Each satellite is represented by three columns: elevation (°), azimuth (°) and range (km).

The settings in `project92` are as follows: frequency 2 GHz, building height 10 m, street width 15 m, separation from the street center 3 m, MS antenna height 1.5 m. The resulting masking angle for the inscribed cylinder is 39.8° (Figure 9.8). Figures 9.10–9.12 illustrate

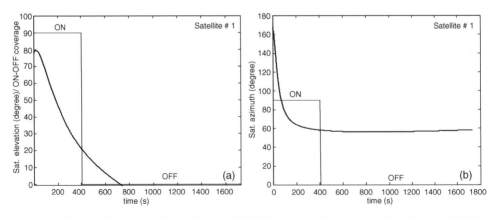

Figure 9.10 (a) Elevation of Sat#1 and ON/OFF time series representing link availability. (b) Azimuth of Sat#1 and ON/OFF time series representing link availability

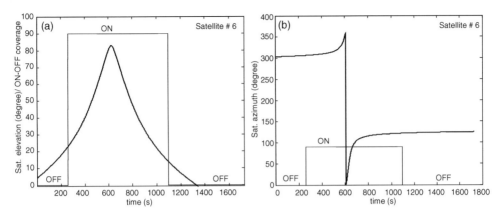

Figure 9.11 (a) Elevation of Sat#6 and ON/OFF time series. (b) Azimuth of Sat#6 and ON/OFF time series

the elevation and azimuth for three out of the 10 satellites contained in `newSatMatrix`. `mat`. Also in these figures, the ON/OFF time series corresponding to the visibility conditions of the link are plotted. Figure 9.13 shows the 10 orbit trajectories on a polar plot and finally Figure 9.14 summarizes the ON/OFF status of all 10 links as a function of time. In this case, it can be observed how, through satellite diversity and handover, it is possible to maintain the availability of the link throughout the simulation period, given that at least one satellite is visible at all times.

Project 9.3: LMS Propagation in Virtual Cities

Here we try to go one step further with respect to Project 2.4 where we concentrated on the modeling of diffraction effects alone. Here we want to introduce other components of the received signal such as the diffuse multipath. In Project 2.4 we represented the buildings on

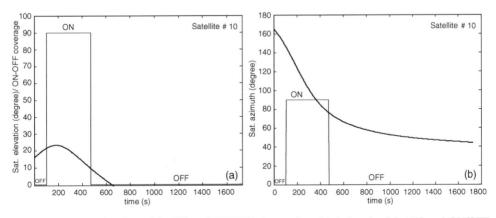

Figure 9.12 (a) Elevation of Sat#10 and ON/OFF time series. (b) Azimuth of Sat#10 and ON/OFF time series

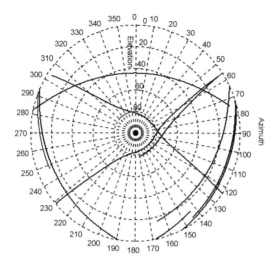

Figure 9.13   Trajectories of all 10 satellites in polar coordinates

the transmitter/satellite side of the street as *rectangular irregularities* or *protuberances* on an infinite screen. The diffraction effects were computed by performing an integration throughout the aperture defined from the screen edge to infinite. The overall integration was split into smaller areas to ease the computation process.

Also, as we have seen in Project 5.4, Rice-distributed time series can be generated by adding two filtered Gaussian random generators in quadrature, representing the diffuse multipath component, to a constant, representing the direct signal.

Here we replicate this process but we use, instead of a constant direct signal, a variable one corresponding to the result of the diffraction calculation. This will give rise to a Ricean distribution with a variable *k-factor* driven by the variations in the diffracted signal. This may not be a totally rigorous procedure but can be considered as a simplified method for generating realistic time series. This procedure can be used as an alternative to the Markov approach presented in Project 9.1. Figure 9.15 illustrates the overall process.

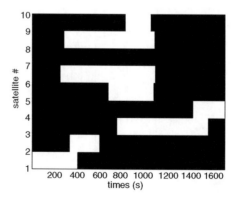

Figure 9.14   Availability state of all 10 satellites as a function of time. White means 'available' and black 'unavailable'

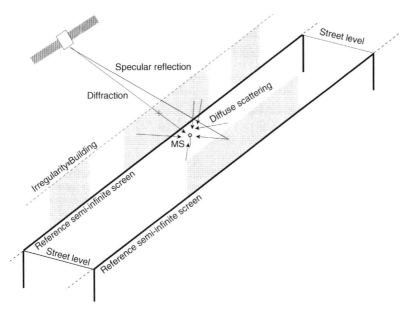

Figure 9.15   Simulation scenario in Project 9.3

The input data to this method can either be a real, specific scenario where an actual building layout is modeled by rectangular irregularities on the infinite screen, or it can be a randomly generated scenario produced using measured distributions of building sizes: virtual city [9] or physical statistical approach [10]. As for the diffuse multipath power levels used in the simulations, these can be drawn from measurements. An alternative would be to use rough surface scattering models [11]. This is beyond the scope of this book.

We have run `project93` with the following settings: frequency 2 GHz, normalized average multipath power −15 dB, mobile speed 10 m/s, sampling fraction of wavelength 4. The input scenario is the same as in Project 2.4 and the Doppler-shaping filter is of the Butterworth type with the same parameters as in Project 5.4.

Figure 9.16 shows the resulting diffracted signal for an elevation angle of 30° and a relative azimuth of 0°. Figure 9.17 shows the generated diffuse multipath. Here, unlike in

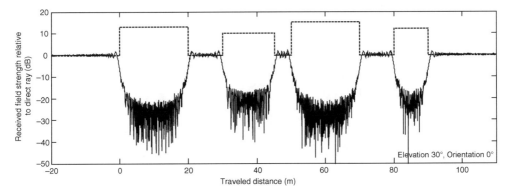

Figure 9.16   Series corresponding to diffraction only

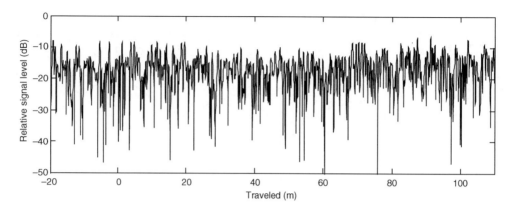

Figure 9.17 Series corresponding to multipath only

Project 9.1, we have assumed the same multipath power for the GOOD and BAD states. This is a reasonable assumption given that this parameter is less significant than the direct signal amplitude, as we have discussed in Project 9.1.

Finally, Figure 9.18 shows the resulting time series after the complex sum of the diffracted and the diffuse signals. The resulting signal represents, in a much more realistic way, what actually takes place. As said above, this approach can be an alternative to a purely statistical method such as that presented in Project 9.1 based on a Markov chain plus Rayleigh/Rice variations within the states.

This approach can be further enhanced if the specular reflection off the building faces on the opposite side of the street is taken into account. The reader is encouraged to implement this simulation but is warned of the fact that the reflection coefficient for circular polarization may change its sense of rotation if the Brewster angle is exceeded (Chapter 8).

Project 9.4: Doppler Shift

Here we discuss the two Doppler-related phenomena present in the LMS channel: the *Doppler shift* and the *Doppler spread*. The Doppler shift is significant for non-GEO satellites

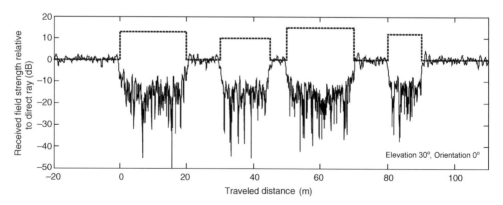

Figure 9.18 Series corresponding to diffraction and multipath

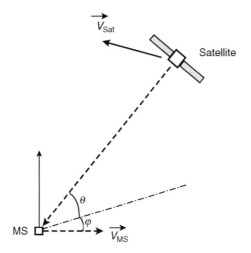

Figure 9.19   Geometry of the Doppler effects

where the relative positions of the satellite and the user change (Figure 9.19). The Doppler shift is proportional to the radial velocity between the satellite and the terminal. These shifts can be modeled separately from those due to the movement of the terminal with respect to the scattering environment: the Doppler spread. The Doppler spread has already been dealt with in the preceding chapters.

Doppler shifts due to satellite motion can be of several kHz, although the variation in this shift is fairly slow, even for LEO satellites at L/S-bands, and it must be tracked and corrected for at the receiver. Figure 9.20 illustrates both the Doppler shift and Doppler spread concepts.

We have performed simulations of the Doppler shift in `project94`, where we have computed the time series of the distance between the satellite and the mobile terminal, assumed to be stationary. We then computed the phase time series, $\phi(t) = (2\pi/\lambda)d(t)$, and finally, its discrete time derivative, i.e., $\Delta\phi(t)/(2\pi\Delta t)$, to compute the Doppler shift.

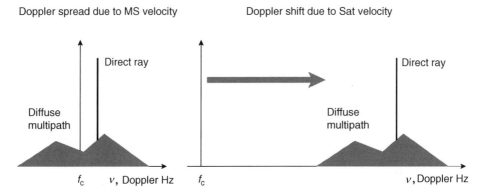

Figure 9.20   Schematic representation of the Doppler spread and Doppler shift phenomena

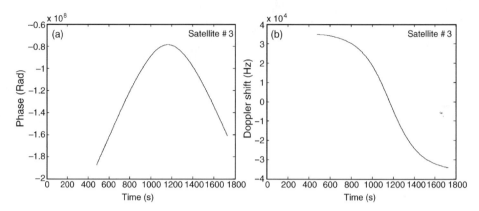

Figure 9.21 (a) Phase time series. (b) Doppler shift time series for Sat#3

Figure 9.21 illustrates the phase and Doppler shift series for Satellite #3 in `newSatMatrix.mat`. Note that Doppler shifts of the order of kHz can appear.

9.3 Summary

In this chapter, we have become acquainted with some of the specific issues of the LMS channel, and its more common modeling techniques. We have presented alternative ways for generating time series, i.e., using a fully statistical approach based on Markov chains, and a mixed statistical–deterministic approach called here the virtual city approach. Additionally, we have become familiar with simple techniques for assessing multiple satellite availability by making use of constellation simulator data together with urban masks. Finally, we have quantified the Doppler shift caused by non-GEO satellites. Next, in the last chapter of this book, we try to get some insight into the spatial characteristics of the multipath channel.

References

[1] C. Loo. A statistical model for land mobile satellite link. *IEEE Trans. Vehic. Tech.*, **34**, 1985, 122–127.

[2] E. Lutz, D.M. Dippold, F. Dolainsky & W. Papke. The land mobile satellite communications channel – Recording, statistics and channel model. *IEEE Trans. Vehic. Tech.*, **40**, 1991, 375–386.

[3] B. Vucetic & J. Du. Channel model and simulation in satellite mobile communication systems. *IEEE Trans. Vehic. Tech.*, **VT-34**, 1985, 122–127.

[4] F.P. Fontan, M. Vazquez-Castro, C. Enjamio, J. Pita García & E. Kubista. Statistical modeling of the LMS channel. *IEEE Trans. Vehic. Tech.*, **50**, 2001, 1549–1567.

[5] H. Suzuki. A statistical model for urban radio propagation. *IEEE Trans. Comm.*, **25**(7), 1979, 213–225.

[6] G.E. Corazza & F. Vatalaro. A statistical model for land mobile satellite channels and its application to non-geostationary orbit systems. *IEEE Trans. Vehic. Tech.*, **43**, 1994, 738–742.

[7] L.E. Bråten & T. Tjelta. Semi-Markov multistate modeling of the land mobile propagation channel for geostationary satellites. *IEEE Trans. Antennas Propag.*, **50**(12), 2002, 1795–1802.

[8] R. Akturan & W. Vogel. Photogrammetric mobile satellite service prediction. *Electronic Lett.*, **31**(3), 1996, 165–166.

[9] Y. Karasawa, K. Minamisoto & T.Matsudo. Propagation channel model for personal mobile satellite systems, PIERS'94 Conference, Noordwijk, The Netherlands.

[10] S.R. Saunders. *Antennas and Propagation for Wireless Communication Systems*, Chapter 9, John Wiley & Sons, Ltd, Chichester, UK, 1999.

[11] P. Beckman & A. Spizzichino. *The Scattering of Electromagnetic Waves from Rough Surfaces*, Macmillan, New York, 1963.

Software Supplied

In this section, we provide a list of functions and scripts, developed in MATLAB®, implementing the various projects mentioned in this chapter. They are the following:

```
project91            filtersignal
project92            rayleigh
project93            mmpolar
project94
```

Also, the satellite constellation time series contained in `newSatMatrix.mat` is provided. Additionally a constellation simulator, `generator`, is made available consisting of the following scripts and functions:

```
generator            CH_LEAP
                     JUL_DATE
                     orbits
                     picksats
                     read_tle
```

where `picksats` was used for generating `newSatMatrix.mat` used in some of the projects. The reader can run other constellations.

Files containing tle data for several well-known constellations are also provided although they might become outdated soon after this book is released. Thus, new tle files for the same constellations can be downloaded from the web, e.g. from http://celestrak.com/NORAD/elements/. The following tle files are available: globalstar.txt, iridium.txt, xmradio.txt, gps-ops.txt, sirius.txt.

10

The Directional Wireless Channel

10.1 Introduction

So far we have only considered the so-called *single-input-single-output* (SISO) channel. Two additional dimensions of the multipath channel are being exploited in new and future systems: space and polarization. Here, the directional channel is modeled and simulated. Techniques such as smart antennas and *multiple-input multiple-output* (MIMO) are attracting the attention of industry, given the advantages that can be drawn: diversity and spatial multiplexing [1] (Figure 10.1). In the figure, the different axes for exploiting the spatial domain are pointed out, i.e., capacity improvement, diversity or array gain. In this chapter, several spatial channel-related simulations are proposed. Polarization aspects have been left out of this book, given the inherent geometrical difficulty in performing vector simulations. Some aspects of the spatial multipath channel have already been treated when we discussed diversity in Chapter 5.

10.2 MIMO Systems

In this section we briefly introduce some MIMO-related concepts that we are going to need in the projects below [2]. By using such systems it is possible to obtain capacities much larger than those obtained using conventional single antenna systems. Work in this respect is ongoing in standardization organizations such as 3GPP dealing with cellular 3G systems, or IEEE 802.11n dealing with wireless LANs.

It is possible to use multiple antennas at only one end of the link. This allows increasing the directivity in a given direction, that of the wanted signal, and creates nulls in other directions, those of the interferers.

The Shannon theorem can be generalized for multiple channels. It provides a quantification of the capacity over a Gaussian communication link when an adequate coding scheme is used, i.e., the maximum bits per second rate achievable without incurring in transmission errors. Thus, the capacity, C, is given by

$$C = \log_2(1 + \rho) \text{ b/s/Hz} \tag{10.1}$$

Modeling the Wireless Propagation Channel   F. Pérez Fontán and P. Mariño Espiñeira
© 2008 John Wiley & Sons, Ltd

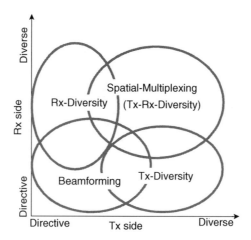

Figure 10.1 Alternatives to exploiting the spatial multipath channel [1]

where ρ is the signal-to-noise ratio. This is a theoretical maximum, unachievable in practice, although it is possible to get closer to it using, for example, turbo codes. If we multiply C by the bandwidth, B (Hz), we get the capacity in bits per second.

For an m-antenna array system, the received signal will be m times larger (m^2 times more power) and the resulting capacity (C_{array}) will be

$$C_{\text{array}} = \log_2(1 + m^2\rho) \ \text{b/s/Hz} \tag{10.2}$$

If we had m equal, independent channels, the combined capacity (C_m) would be

$$C_m = m \, \log_2(1 + \rho) \ \text{b/s/Hz} \tag{10.3}$$

Comparing the above expressions, we can say that it is possible to achieve a much larger capacity with m independent radio links than with an array of m antennas. Now we try to look into how this can be achieved. For this, we first introduce the concept of the MIMO *transmission matrix*. For the SISO case, the relationship between the input and output in the frequency domain can be written as

$$y(f) = h(f) \cdot x(f) \tag{10.4}$$

where $x(f)$ and $y(f)$ are the transmitted and received signals, and $h(f)$ is the channel transfer function in the low-pass equivalent domain (Figure 10.2), i.e., function $T(f, t)$ in the wideband channel terminology of Chapter 7. In the following, we drop the f variable, as we will be assuming that the channel is narrowband. The notation may seem misleading but, basically, when we speak about h at a given frequency, we are referring to the channel's time-varying complex envelope, which we have been calling $r(t)$ up to now, that is, we particularize $T(f, t)$ for a given frequency, e.g., f_c, and then, $T(f_c, t) = r(t)$. However, we will continue using h, as it is the most common notation used in papers dealing with MIMO systems.

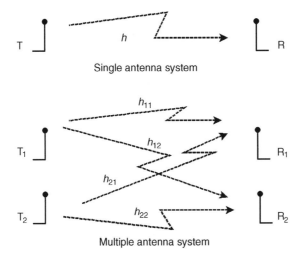

Figure 10.2   Propagation between two antenna pairs

As shown in Figure 10.2, in a multiple antenna system, the propagation conditions can be different between each transmit and receive antenna. To characterize the system, a single channel function, h, such as that in Equation 10.4 is not sufficient, we need a set of as many channel functions as possible Tx–Rx antenna pairs, in the case of Figure 10.2, we need 2×2 [2],

$$y_1 = h_{11}x_1 + h_{12}x_2$$
$$y_2 = h_{21}x_1 + h_{22}x_2$$
(10.5)

where x_i and y_j are the input and output complex voltages at transmit antenna i and receive antenna j, respectively, and h_{ij} is the complex transfer function between transmit antenna j and receive antenna i. Putting the above equation in matrix form we get

$$\begin{bmatrix} y_1 \\ y_2 \end{bmatrix} = \begin{bmatrix} h_{11} & h_{12} \\ h_{21} & h_{22} \end{bmatrix} \begin{bmatrix} x_1 \\ x_2 \end{bmatrix} \quad \text{or} \quad \mathbf{y} = \mathbf{H} \cdot \mathbf{x}$$
(10.6)

where $\mathbf{y}$ and $\mathbf{x}$ are the output and input complex voltage vectors, and $\mathbf{H}$ the *transmission matrix*.

The capacity will be maximum when the channels are independent. This will occur if matrix $\mathbf{H}$ is diagonal, i.e., $\mathbf{H} = \mathbf{D}$, that is, when each antenna receives only from a single transmit antenna, i.e.,

$$\begin{bmatrix} y_1 \\ y_2 \end{bmatrix} = \begin{bmatrix} d_1 & 0 \\ 0 & d_2 \end{bmatrix} \begin{bmatrix} x_1 \\ x_2 \end{bmatrix} + \begin{bmatrix} n_1 \\ n_2 \end{bmatrix} \quad \text{or} \quad \mathbf{y} = \mathbf{D} \cdot \mathbf{x} + \mathbf{n}$$
(10.7)

where a noise vector, $\mathbf{n}$, has been introduced for completeness. Such conditions hardly occur in reality. What we need is to 'diagonalize' matrix $\mathbf{H}$. This is equivalent to performing some

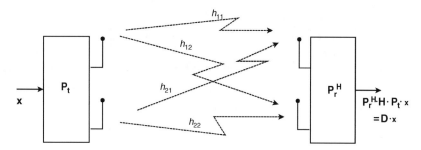

Figure 10.3   MIMO processing for obtaining independent propagation paths

sort of signal processing at both the transmit and receive sides, which converts the channel plus signal processing ensemble into a diagonal matrix. This can be achieved by introducing at both ends two matrices, $\mathbf{P_r}$ and $\mathbf{P_t}$, such that

$$\mathbf{P_r^H} \cdot \mathbf{H} \cdot \mathbf{P_t} = \mathbf{D} \tag{10.8}$$

where ^{H} means transposed hermitic matrix (conjugate transposed). The overall transmission plus signal processing chain is given by

$$\mathbf{y} = \mathbf{P_r^H} \cdot \mathbf{H} \cdot \mathbf{P_t} \cdot \mathbf{x} = \mathbf{D} \cdot \mathbf{x} \tag{10.9}$$

The result (Figure 10.3) is that we obtain independent propagation paths. Of course, this processing has to be dynamic as the channel is time varying.

The 'diagonalization' process can be achieved by performing the *singular value decomposition* (SVD) of $\mathbf{H}$. Assume we have a complex $m \times m$ matrix $\mathbf{H}$, and we want to put it in the form

$$\mathbf{H} = \mathbf{U} \cdot \mathbf{D} \cdot \mathbf{V^H} \tag{10.10}$$

with $\mathbf{D}$ being a diagonal matrix, and $\mathbf{U}$ and $\mathbf{V}$ unitary *orthonormal matrices*, thus

$$\mathbf{U} \cdot \mathbf{U^H} = \mathbf{I} \text{ and } \mathbf{V} \cdot \mathbf{V^H} = \mathbf{I}, \tag{10.11}$$

where $\mathbf{I}$ is the $m \times m$ *identity matrix*. By performing the operation

$$\mathbf{D} = \mathbf{U^H} \cdot \mathbf{H} \cdot \mathbf{V} \tag{10.12}$$

we arrive at the signal processing we required for performing the transmission matrix 'diagonalization'.

The elements, d_i, of the diagonal matrix, $\mathbf{D}$, are also the square roots of the eigenvalues of $\mathbf{H} \cdot \mathbf{H^H}$ and $\mathbf{H^H} \cdot \mathbf{H}$. These two matrices are *hermitic*, and thus their eigenvalues are either positive or zero. The *rank*, r ($r \leq m$), is the number of non-zero eigenvalues.

The columns, $\mathbf{U}_i$, of matrix $\mathbf{U}$ are *orthonormal eigenvectors* of $\mathbf{H} \cdot \mathbf{H^H}$. Similarly columns, $\mathbf{V}_i$, of matrix $\mathbf{V}$ are *orthonormal eigenvectors* of $\mathbf{H^H} \cdot \mathbf{H}$.

For example [2], assuming a transmission matrix such that

$$\mathbf{H} = \begin{bmatrix} 0.5 & 0.3 \\ -j0.6 & j0.4 \end{bmatrix} \tag{10.13}$$

Its hermitic transposed, $\mathbf{H^H}$, and the product $\mathbf{H^H} \cdot \mathbf{H}$ are

$$\mathbf{H^H} = \begin{bmatrix} 0.5 & j0.6 \\ 0.3 & -j0.4 \end{bmatrix} \quad \text{and} \quad \mathbf{H^H} \cdot \mathbf{H} = \begin{bmatrix} 0.61 & -0.09 \\ -0.09 & 0.25 \end{bmatrix} \tag{10.14}$$

The *eigenvalues* of $\mathbf{H^H} \cdot \mathbf{H}$ are $d_1^2 = 0.6312$ and $d_2^2 = 0.228$, which can be calculated using MATLAB® (MATLAB® is a registered trademark of The MathWorks, Inc.) function eig. The elements of the diagonal matrix $\mathbf{D}$, d_1 and d_2, can also be calculated using MATLAB® function svd, i.e,

$$\mathbf{D} = \begin{bmatrix} 0.7945 & 0 \\ 0 & 0.4783 \end{bmatrix} \tag{10.15}$$

We reproduce below results from the MATLAB® workspace:

```
H =
  0.5000        0.3000
  0 - 0.6000i    0 + 0.4000i

>> [U,S,V] = svd(H)

U =
  -0.5257      -0.8507
   0 + 0.8507i  0 - 0.5257i

S =
  0.7945        0
       0      0.4783

V =
  -0.9732      -0.2298
   0.2298      -0.9732

>> HH=H'*H

HH =
   0.6100      -0.0900
  -0.0900       0.2500

>> [VV,DD] = eig(HH)
```

```
VV =
   -0.2298      -0.9732
   -0.9732       0.2298

DD =
    0.2288           0
         0      0.6312
```

Having computed matrices $\mathbf{P_r}$ and $\mathbf{P_t}$, it is possible to obtain as many independent output signals as pairs of antennas there are, i.e.,

$$y_i = d_i \cdot x_i \quad \text{with } i = 1, 2, \ldots, m \tag{10.16}$$

The amount of power transferred between x_i and y_i is given by $|d_i|^2$. Channel i is called *propagation mode i*. There can be as many possible modes as antenna pairs. The greater coefficient d_i is, the larger its signal-to-noise ratio becomes and hence the capacity achievable through mode i.

To illustrate the importance of the values of d_i^2 and their influence on the mode capacity, we can look at different examples represented by their transmission matrices [2]. A first case would correspond to line-of-sight propagation conditions where all paths are identical. The corresponding transmission and diagonal matrices are given for example by

$$\mathbf{H} = \begin{bmatrix} 1 & 1 \\ 1 & 1 \end{bmatrix} \quad \text{and} \quad \mathbf{D} = \begin{bmatrix} 2 & 0 \\ 0 & 0 \end{bmatrix} \tag{10.17}$$

and thus $|d_1|^2 = 4$, $|d_2|^2 = 0$. In the second case, we assume that the differences between antennas are very small, e.g.,

$$\mathbf{H} = \begin{bmatrix} 1 & 0.98 \\ 0.99 & 1 \end{bmatrix} \quad \text{and} \quad \mathbf{D} = \begin{bmatrix} 1.985 & 0 \\ 0 & 0.015 \end{bmatrix} \tag{10.18}$$

and thus $|d_1|^2 = 3.9403$, $|d_2|^2 = 0.0002$. Finally, the third case shows significant differences where, for example,

$$\mathbf{H} = \begin{bmatrix} 0.5 & 0.3 \\ -j0.6 & j0.4 \end{bmatrix} \quad \text{and} \quad \mathbf{D} = \begin{bmatrix} 0.7945 & 0 \\ 0 & 0.4783 \end{bmatrix} \tag{10.19}$$

and thus $|d_1|^2 = 0.6312$, $|d_2|^2 = 0.2288$. This would correspond to a *multipath-rich* propagation scenario.

In the first case, propagation is only through the direct ray. This means that all the power is transmitted through one antenna pair (d_1), while the other pair does not contribute any power $(d_2 = 0)$. As propagation differences appear, the relevance of the power at the output of the first receive antenna diminishes while there starts to be power in the second. The following conclusions can be drawn [2]:

- When the direct LOS signal dominates, MIMO systems do not provide any practical advantage. Only one mode is excited.

- MIMO systems are advantageous when differences exist in the propagation conditions between the various antenna pairs. In this case, significant modes can be excited. For this to happen the following conditions are necessary: (a) both the transmit and receive antennas must be sufficiently spaced; and (b) multipath is the dominating propagation mechanism, with a very small direct component.

Now, we come back to the issue of capacity [3]. For completeness, we review some single and multiple antenna configurations. For a 1×1 (SISO) system the capacity is given by

$$C_{\text{SISO}} = \log_2(1 + \rho|h|^2) \text{ b/s/Hz} \tag{10.20}$$

where h is the normalized complex gain of a particular realization of a random channel. A second configuration of interest is that of a *single-input-multiple-output* (SIMO) system with M Rx (*receive diversity*). In this case the capacity is [3]

$$C_{\text{SIMO}} = \log_2\left(1 + \rho \sum_{i=1}^{M} |h_i|^2\right) \text{ b/s/Hz} \tag{10.21}$$

where h_i is the channel gain for Rx antenna i.

Another option is using *transmit diversity*, with the transmitter not knowing the channel state. This is a *multiple-input-single-output* (MISO) system with N Tx antennas. In this case, the capacity is [3]

$$C_{\text{MISO}} = \log_2\left(1 + \frac{\rho}{N} \sum_{i=1}^{N} |h_i|^2\right) \text{ b/s/Hz} \tag{10.22}$$

where the transmit power is equally split between the N transmit antennas.

Finally, for the MIMO case with N Tx and M Rx antennas, the capacity is given by [3]

$$C_{\text{MIMO}} = \log_2\left[\det\left(\mathbf{I_M} + \frac{\rho}{N}\mathbf{HH}^*\right)\right] \text{ b/s/Hz} \tag{10.23}$$

where (*) means transpose-conjugate and $\mathbf{H}$ is the $M \times N$ channel transmission matrix. This equation can be rewritten as

$$C_{\text{MIMO}} = \sum_{i=1}^{m} \log_2\left(1 + \frac{\rho}{N}d_i^2\right) \text{ b/s/Hz} \tag{10.24}$$

where $m = \min(M,N)$ and the terms d_i have already been defined above. We can see the significant capacity enhancement brought about by the MIMO configuration over the others, as the summation operator is now outside the logarithm operator. Of course, it is pointless wasting power on those modes where $|d_i|$ is zero or negligible. It is advisable to apply most of the power to those modes with the largest $|d_i|$ values. The calculation of the optimal share of power leads to so-called *water-filling* techniques.

We are going to use the above equations in Project 10.2 for computing the capacity for the MIMO and SISO configurations.

One important practical implementation problem is that MIMO systems require knowledge of the channel conditions through *sounding techniques* which should provide the current values of **H** in real time. As we know, the rate of change of the channel is given by its *coherence time*. For stationary terminals, e.g., in WLANs, the channel variability will be small and normally due to people moving about in the vicinity of either end of the link.

To perform *channel sounding*, the transmit signal must include a known sequence or preamble. Each antenna has to have an associated sequence, different from the others. The channel sounding allows the receiver to know the current channel's state. In some implementations, it is also necessary that the transmitter knows what the current channel state is – this requires the setting up of a *return channel*. This may impose on the system a great burden. It is also common to employ so-called *bind techniques* at the expense of more complex signal processing.

10.3 Projects

After this brief introduction, we propose here a number of simulation projects that try to present some of the basic issues relative to the directional channel including the clustering of rays, angles of arrival and departure, and MIMO systems.

Project 10.1: Generating Clustered Point-Scatterer Propagation Scenarios

In order to generate synthetic scenarios made up of multiple point scatterers, we apply two enhanced versions of the *Saleh and Valenzuela model* (S&V) [4], which was discussed in Chapter 8. There we concentrated on generating time-varying impulse responses, $h(t,\tau)$, and tapped delay lines. In this project, we will be using enhancements to the S&V model due to Spencer *et al.* [5] and Wallace and Jensen [6], where assumptions on the spatial distribution of ray clusters are made for the SIMO/MISO and narrowband MIMO cases, respectively. The spatial enhancements proposed deal with the *angle of arrival* (AoA) in [5], and both the *angle of departure* (AoD) and the AoA in [6].

From measurements carried out in indoor environments, it was found [5] that ray clusters are not only spread out in delay but also in AoA. Thus, the original S&V model was extended to include angles of arrival, θ, as indicated by the equation

$$h(\tau,\theta) = \sum_l \sum_k \beta_{kl} \exp(j\phi_{kl})\delta(\tau - T_l - \tau_{kl})\delta(\theta - \Theta_l - \theta_{kl}) \qquad (10.25)$$

where index l indicates the cluster number and index kl means ray k within cluster l. Only a horizontal 2D distribution is assumed here. The model proposed in [5] for the angles of arrival, θ, is a zero-mean *Laplace distribution* with standard deviation σ_P,

$$p(\theta_{kl}) = \frac{1}{\sqrt{2}\sigma_P}\exp(-|\sqrt{2}\theta_{kl}/\sigma_P|) \qquad (10.26)$$

and where Θ_l is *uniformly distributed* between 0 and 2π. Θ_l is the mean AoA for cluster l while θ_{kl} is an angle relative to the mean.

An extension to [5] was proposed in [6] for the narrowband case, which includes both AoAs and AoDs. The new expression is then

$$h(\tau,\theta^T,\theta^R) = \sum_l \sum_k \beta_{kl}\exp(j\phi_{kl})\delta(\tau - T_l - \tau_{kl})\delta(\theta^T - \Theta_l^T - \theta_{kl}^T)\delta(\theta^T - \Theta_l^R - \theta_{kl}^R) \quad (10.27)$$

where we have kept the delay variable. In this expression, the coefficients β_{kl} are Rayleigh distributed. The same statistical assumptions are made for the angles of arrival and departure, i.e., uniform cluster angles and Laplacian ray angles.

The main objective of this project is creating synthetic multiple point-scatterer scenarios based on the above models. One difference in interpretation is important though: we are going to assume, in this case, that the coefficients β_{kl} are path-length dependent but constant in magnitude. The Rayleigh fade phenomenon will appear through the complex addition of rays at the receiver as this moves.

We will be running within `project101` the S&V model implementation developed for `project84`, but we will include the new features just discussed.

For modeling the excess delays, BS and MS are assumed to be located on either foci (F_1 and F_2) of a set of ellipses. Their separation is $2d$. For each possible excess delay, a different ellipse is defined. As discussed in Chapter 7, scatterers giving rise to the same excess delay will be located on the same ellipse. The general equation of an ellipse (Figure 10.4) is given by

$$\frac{x^2}{a^2} + \frac{y^2}{b^2} = 1 \quad (10.28)$$

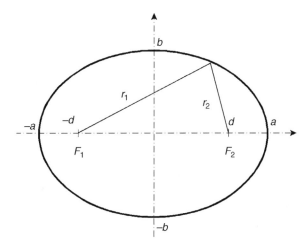

Figure 10.4 Geometry and parameters of an ellipse

with its parameters a and b (semi-major and semi-minor axes) following the expression, $a^2 - d^2 = b^2$. Other expressions of interest are as follows:

$$\varepsilon = \frac{d}{a} = \frac{\sqrt{a^2 - b^2}}{a} < 1, \quad r_1 = a + \varepsilon x \quad \text{and} \quad r_2 = a - \varepsilon x \tag{10.29}$$

where ε is called the eccentricity.

The radio path of the direct ray, which we assume completely blocked, is $2d$; while the radio path, by way of reflection on a point scatterer, on one of the ellipses, is given by the sum of its corresponding sub-paths, i.e., $r_1 + r_2$. Thus, the *excess path length* and its associated *excess delay* are given by:

$$\Delta r = r_1 + r_2 - 2d = a + \varepsilon x + a - \varepsilon x - 2d = 2(a - d)$$
$$\text{and} \quad \Delta \tau = \Delta r / c = 2(a - d)/c \tag{10.30}$$

where c is the speed of light. For a given excess delay, $\Delta \tau_i$, knowing the BS–MS separation, $2d$, we need to compute the semi-major, a, and semi-minor, b, axes of the corresponding ellipse. Thus,

$$a_i = \frac{c \Delta \tau_i}{2} + d \quad \text{and} \quad b_i = \sqrt{a_i^2 - d^2} \tag{10.31}$$

For generating the nominal position of cluster i, we need to draw values for both Θ_i^T and Θ_i^R using a uniform distribution (MATLAB® function rand). We are going to draw only the value for Θ_i^R, and relax the requirement of a uniform distribution for Θ_i^T, which we will calculate geometrically from the excess delay, $\Delta \tau_i$, and Θ_i^R. Thus, once a Θ_i^R value has been randomly drawn, we have to calculate the nominal position of the scatterer on ellipse i. Thus,

$$\cos(\Theta_i^R) = \frac{x_i - d}{r_2} = \frac{x_i - d}{a_i - \varepsilon x_i} \tag{10.32}$$

consequently,

$$x_i = \frac{d + a_i \cos(\Theta_i^R)}{1 + \varepsilon \cos(\Theta_i^R)} \quad \text{and} \quad y_i = (a_i - \varepsilon x_i) \sin(\Theta_i^R) \tag{10.33}$$

In project101 we have located the origin of coordinates on MS. This means that a translation of x_i by $-d$ m must be carried out (Figure 10.5) with respect to Figure 10.4.

The actual point-scatterer angles are Laplace distributed. We have to draw Laplace distributed angles with respect to each of the cluster arrival angles, Θ_i^R. This random generator was built using the inverse CDF technique illustrated in Figure 10.6. We drew uniform random samples on the probability axis (ordinates) and, by reading the corresponding abscissa, we were able to generate a Laplace-distributed series (Function genLaplacian).

The settings for project101 were as follows: the frequency was 1500 MHz, the distance between BS and MS 50 m, the propagation law exponent was $n = 4$, the EIRP

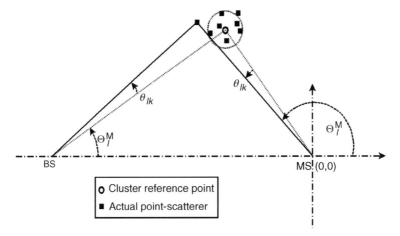

Figure 10.5 Geometrical parameters for generating cluster reference points and actual point-scatterer locations within `project101`

1 W, the Laplace parameter $\sigma_P = 2.5$, the S&V model parameters were $\lambda = 5$, $\Lambda = 25$, $\gamma = 29$ and $\Gamma = 60$.

Figure 10.7(a) shows the resulting point-scatterer scenario where clusters have been created through the algorithms outlined above following the S&V model, and its spatial enhancements. Figure 10.7(b) shows the corresponding *scattering function* where delays and

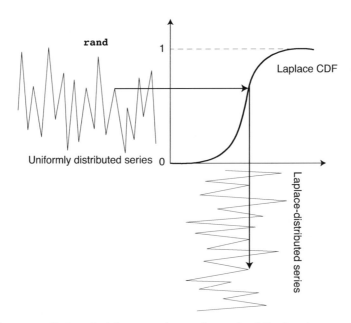

Figure 10.6 Inverse CDF method for generating random series following the wanted distribution

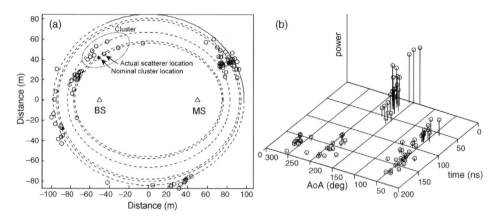

Figure 10.7 (a) Synthesized point-scatterer scenario based on the S&V model and its spatial enhancements. (b) Scattering function: delays and angles of arrival corresponding to the synthesized scenario in (a)

angles of arrival are shown. This methodology allows the generation of realistic spatial distributions of multiple point scatterers, which can be used for simulating the channel behavior. Next, in Project 10.2 we apply the multiple point-scatterer model for simulating a number of MIMO-related parameters.

It is important at this point to indicate that the underlying model we are assuming is a single scattering one (single bounce). This means that, even though double scattering might take place, the received power of such contributions would be negligible with respect to that of single interactions. This approximation is realistic in most cases, but there could be some instances, especially when large scatterers or specular reflectors are present, that significant double interactions could arise.

Project 10.2: MIMO Channel Modeling Using the Multiple Point Scatterers

In this project, we performed several simulations assuming a 3×3 MIMO configuration using the multiple point-scatterer model used throughout this book. The simulation approach is exactly the same as that used in `project511`, the only difference being that now we consider several possible transmitters and receivers with a given spacing between them. One important simplification that we made is to only consider the *narrowband case*.

The basic simulation step is calculating the complex envelope time series for each possible combination of input and output antennas. The ensemble of complex envelopes constitutes the channel transmission matrix. From there, we perform the singular value decomposition of **H** for assessing the available capacity. In the first two simulations (`project1021` and `project1022`) we have assumed that the channel variations occur due to the fact that MS moves at a given speed, V, and thus traverses a spatial standing wave. However, in `project1023` we have considered different realizations of a static case, where the only thing that changes is the phase term in each scatterer contribution, which is randomly drawn each time.

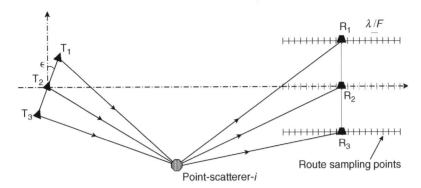

Figure 10.8   Geometry assumed in `project1021`, `project1022` and `project1023`

As we have already seen in Chapter 4, the general representation of the complex envelope received at antenna j from transmit antenna i is given by,

$$r^{ij}[n] = \sum_{i=1}^{N} a_k \exp(-\mathrm{j}k_\mathrm{c}d_k^{ij}[n] + \mathrm{j}\phi_k^{ij}[n]) \tag{10.34}$$

where $[n]$ indicates either sample point along the MS route or, alternatively, realization number. Coefficients ij indicate Tx–Rx antenna pair, k indicates the scatterer number. From the equation, we have assumed a constant scatterer amplitude contribution. However, the phase is different for each antenna pair.

Figure 10.8 illustrates the model geometry where a single point scatterer is shown. In most projects, including `project1021` and `project1022`, we have neglected the phase term, as the phase varied due to the changes in the receiver position. For the simulations performed with `project1023`, for each realization which produced nine (3 × 3) complex values, one per antenna pair, we drew one phase value per scatterer, k, and antenna pair, ij. Thus, as we assumed 100 point scatterers (`NSC`) and had nine antenna pairs, we drew in each realization 900 phase terms.

In the case of `project1021` we selected the following settings: frequency 2 GHz, MS speed 10 m/s, eight samples per wavelength of traveled route, number of scatterers 100, Rayleigh parameter $2\sigma^2$ equal to 0 dB. The spacing between antennas is one wavelength at both ends, and the reference signal-to-noise ratio was `SNR=20` dB. Angle ε was set to 0° (Figure 10.8).

Figure 10.9 plots the actual multiple point-scatterer scenario simulated in `project1021` where a circle of scatterers was assumed about MS. For convenience, we have opted here for a circle-of-scatterers scenario. This scenario was already used in Chapter 5 for assessing diversity gains at the BS and MS sides. It is left for the reader to implement the linkage between the scenarios generated in Project 10.1 and the simulators in this project.

Figure 10.10 illustrates the nine time series corresponding to all possible antenna pairs. Figure 10.11(a) illustrates the time series of the three singular values obtained in the diagonalization process while Figure 10.11(b) shows their CDFs. Figure 10.12(a) shows the

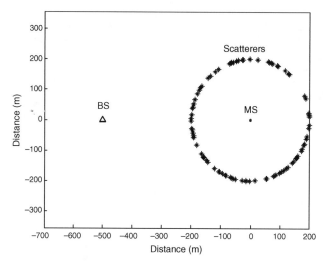

Figure 10.9   Simulated scenario

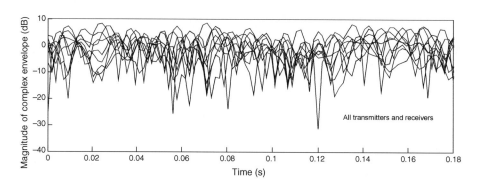

Figure 10.10   Time series for all Tx–Rx antenna pairs

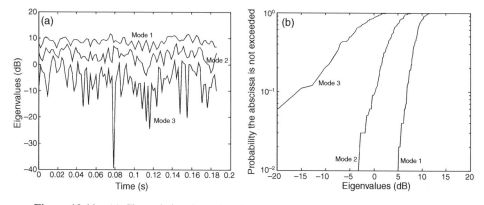

Figure 10.11   (a) Channel singular value time series. (b) Channel singular value CDFs

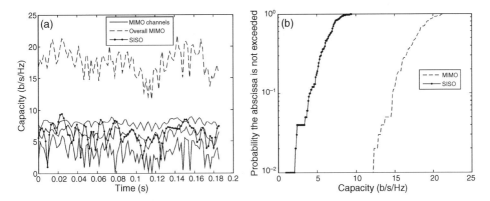

Figure 10.12 (a) Instantaneous capacity time series for SISO, individual MIMO modes and overall MIMO cases. (b) Capacity CDFs for SISO and MIMO cases

time-varying capacities of the individual MIMO modes, the SIMO and the MIMO capacities. It can clearly be observed how MIMO brings about a drastic increase in capacity. Figure 10.12(b) shows the SIMO and MIMO capacity CDFs.

In `project1022` we simulated the same scenario, the only difference is that now the antenna spacings both at BS and MS was $\lambda/16$. Figure 10.13 illustrates the nine resulting time series. As it can be observed, the correlation between the series is very high in comparison with Figure 10.10 corresponding to a one- wavelength antenna separation. Figure 10.14(a) shows the resulting singular values where there is only one of significance, the others being negligible. This is again shown in Figure 10.14(b) where their CDFs are plotted. Finally Figure 10.15 shows the instantaneous capacities and their CDFs for the SIMO and MIMO cases.

Similarly, in `project1023` we have run the same simulation as in `project1021`, except for the fact that MS was stationary. We have produced different channel realizations by making different phase draws as discussed above. The antenna spacings were of one

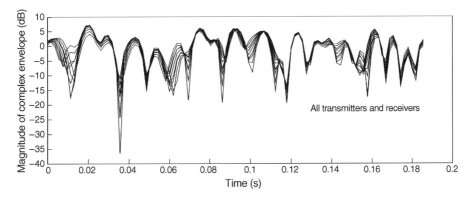

Figure 10.13 Time series for all Tx–Rx antenna pairs

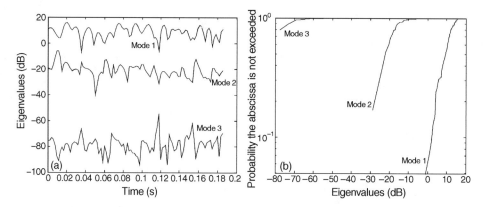

Figure 10.14   (a) Channel singular value time series. (b) Channel singular value CDFs

wavelength. Figure 10.16 illustrates the nine resulting series. Note that the abscissas are realization numbers not time. As it can be observed, the *cross-correlation* between the series is low as in `project1021`. Also note that there should be no *autocorrelation* in the individual series, given that the plotted values correspond to independent realizations. Figure 10.17(a) shows the resulting singular values, again with significant values in all three of them. This is again shown in Figure 10.17(b) where their CDFs are plotted. Finally Figure 10.18 shows the capacities corresponding to each realization and their CDFs for the SIMO and MIMO cases.

Project 10.3: Statistical Modeling of the MIMO Channel

In the previous project, we have shown how using the multiple point-scatterer model we could generate series corresponding to the transmission matrix, **H**, for the narrowband

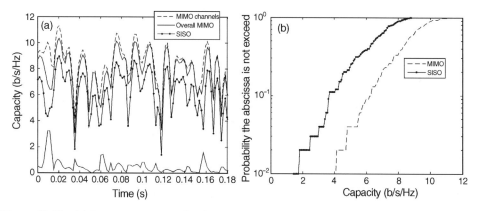

Figure 10.15   (a) Instantaneous capacity time series for SISO, individual MIMO modes and overall MIMO cases. (b) Capacity CDFs for SISO and MIMO cases

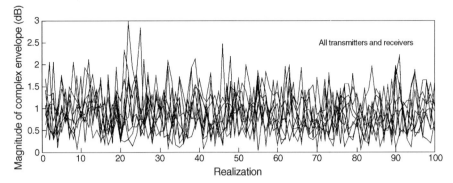

Figure 10.16   Series for all Tx–Rx antenna pairs

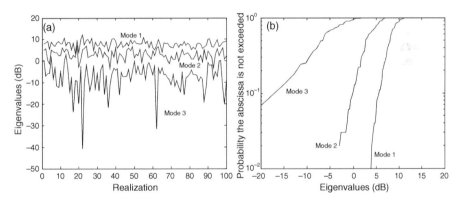

Figure 10.17   (a) Channel singular value time series. (b) Channel singular value CDFs

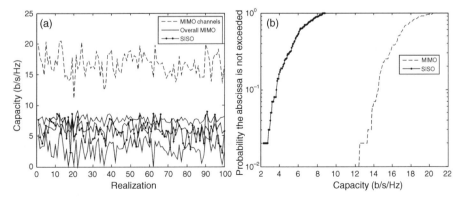

Figure 10.18   (a) Capacity series for SISO, individual MIMO modes and overall MIMO cases. (b) Capacity CDFs for SISO and MIMO cases

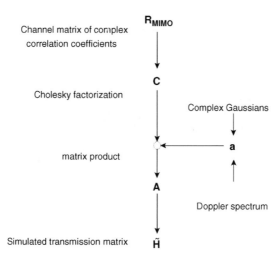

Figure 10.19   Steps for generating the channel transfer matrix elements

MIMO channel. Here, we try another approach based on the correlation matrix, $R_{\mathbf{MIMO}}$. We will be using this matrix from the simulations carried out in the previous project. This matrix was stored in file RMIMO.mat.

The approach implemented in project103 is a statistical one where we initially generate as many time series as possible antenna pairs. These series are totally uncorrelated, thus, their cross-correlations must be forced on the resulting time series, members of **H**, according to the specifications set by matrix $R_{\mathbf{MIMO}}$. Figure 10.19 illustrates the overall process [7] where we have assumed identical Rayleigh statistics for all elements of **H**.

As indicated in the figure, project103 starts from an $M \times N$ $R_{\mathbf{MIMO}}$ matrix calculated previously. The first step is converting the matrix to an $MN \times 1$ vector and then performing its Cholesky factorization to get lower triangular matrix **C**, where $R_{\mathbf{MIMO}} = \mathbf{CC^T}$. This is performed using MATLAB® function chol. On the other hand, as many zero-mean, complex, independent, identically distributed, random variables, a_{mn}, as antenna pairs are generated. Then, they are Doppler shaped using a Butterworth filter, as we did in Project 5.4. This introduces the wanted autocorrelation in the individual series. Matrix **a** will also be an $MN \times 1$ vector. Finally the two matrices are multiplied to obtain the simulated elements of the channel transmission matrix, $\tilde{\mathbf{H}}$, vector **A** which, finally, has to be converted into an $M \times N$ matrix. The multiplication by **C** forces the wanted cross-correlation properties between the series.

The settings for project54 were as follows: frequency 2 GHz, eight samples per wavelength, mobile speed 10 m/s. The Butterworth filter settings were: Wp=0.5*2* fm/fs, Ws=1*2*fm/fs, Rp=3 %dB and Rs=40 %dB. Figure 10.20 shows the resulting time series for all possible antenna pairs. The same parameter calculations carried out in Project 10.2 can be performed for the series generated here. This is left for the reader to implement.

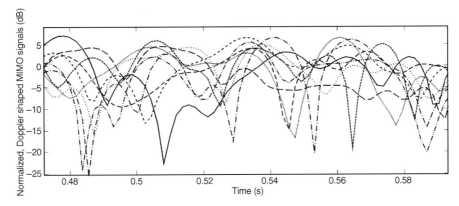

Figure 10.20   Normalized, Doppler-shaped MIMO signals

10.4 Summary

With this chapter, we conclude our study on the wireless channel. Here, we have dealt with the spatial properties of the multipath channel. We have first learned about how scatterer contributions tend to be clustered in terms of excess delays, which indicate that they can belong to the same obstacle. We have also seen that these clustered contributions are spread out in angle of arrival and departure. We have also simulated the MIMO channel using the multiple point-scatterer approach, and we have been able to show how the capacity can be increased substantially. Finally, we have presented another approach, a statistical one, for simulating the directional channel.

There are many other simulations that could have been proposed to the reader. It is hoped, however, that the available simulations can be used as a starting point to becoming familiar with most of the channel features discussed in this book.

References

[1] E. Bonek. The MIMO radio channel. http://publik.tuwien.ac.at/files/pub-et_11093.pdf.

[2] W. Warzanskyj García, L. Miguel Campoy Cervera & J.M. Vázquez Burgos. MIMO antenna arrays: a promise for increasing capacity in mobile communications (in Spanish). *Comunicaciones de Telefónica I+D*, **36**, 2005, 21–34.

[3] D. Gesbert, M. Shafi, D. Shiu, P.J. Smith & A. Naguib. From theory to practice: an overview of MIMO space-time coded wireless systems. *IEEE J. Select. Areas Comm.*, **21**(3), 2003, 281–302.

[4] A.A.M. Saleh & R.A. Valenzuela. A statistical model for indoor multipath propagation. *IEEE J. Select. Areas Comm.*, **SAC-5**(2), 1987, 128–137.

[5] Q. Spencer, B. Jeffs, M. Jensen & A. Swindlehurst. Modeling the statistical time and angle of arrival characteristics of an indoor multipath channel. *IEEE J. Select. Areas Comm.*, **18**(3), 2000, 347–360.

[6] J.W. Wallace & M.A. Jensen. Statistical characteristics of measured mimo wireless channel data and comparison to conventional models. IEEE VTS 54th Vehicular Technology Conference, 2001. VTC 2001 Fall, **2**, 2001, 1078–1082.

[7] J.P. Kermoal, L. Schumacher, K.I. Pedersen, P.E. Mogensen & F. Frederiksen. A stochastic MIMO radio channel model with experimental validation. *IEEE J. Select. Areas Comm.*, **20**(6), 2002, 1211–1226.

Software Supplied

In this section, we provide a list of functions and scripts, developed in MATLAB®, implementing the various projects mentioned in this chapter. They are the following:

```
project101              fCDF
project1021             filtersignal
project1022             genExponential
project1023             genLaplacian
project103              stem2D
```

Index

Printed and bound by CPI Group (UK) Ltd, Croydon, CR0 4YY

17/04/2025

14658871-0001